FRAMING THE INTERPRETER

"*Framing the Interpreter* is a ground-breaking volume that illuminates the power of visual media to shape our understanding of interpreters and interpreter mediation in situations of conflict and beyond. A rich source of conceptual tools and methodological guidance, this work will appeal to scholars working at the interface of disciplines."

Rebecca Tipton, *University of Manchester, UK*

Situations of conflict offer special insights into the history of the interpreter figure, and specifically the part played in that history by photographic representations of interpreters.

This book analyses photo postcards, snapshots and press photos from several historical periods of conflict, associated with different photographic technologies and habits of image consumption: the colonial period, the First and Second World Wars and the Cold War. The book's methodological approach to the "framing" of the interpreter uses tools taken primarily from visual anthropology, sociology and visual syntax to analyse the imagery of the modern era of interpreting. By means of these interpretative frames, the contributions suggest that each culture, subculture or social group constructed its own representation of the interpreter figure through photography.

The volume breaks new ground for image-based research in translation studies by examining photographic representations that reveal the interpreter as a socially constructed category. It locates the interpreter's mediating efforts at the core of the human sciences.

This book will be of interest to researchers and advanced students in translation and interpreting studies, as well as to those working in visual studies, photography, anthropology and military/conflict studies.

Anxo Fernández-Ocampo is Senior Lecturer at the Faculty of Philology and Translation, University of Vigo, Spain.

Michaela Wolf is Associate Professor at the Department of Translation Studies, University of Graz, Austria.

FRAMING THE INTERPRETER

Towards a visual perspective

Edited by
Anxo Fernández-Ocampo and Michaela Wolf

Routledge
Taylor & Francis Group
LONDON AND NEW YORK

First published 2014
by Routledge
4 Park Square, Milton Park, Abingdon, Oxon OX14 4RN
605 Third Avenue, New York, NY 10017

Routledge is an imprint of the Taylor & Francis Group, an informa business

British Library Cataloguing-in-Publication Data
A catalogue record for this book is available from the British Library

Library of Congress Cataloging in Publication Data
Framing the interpreter : towards a visual perspective / edited by
Anxo Fernández-Ocampo and Michaela Wolf.
p. cm
Includes bibliographical references and index.
1. Multimedia systems–Research. 2. Translating and interpreting–Technological innovations. 3. Audio-visual equipment–Technological innovations.
4. Photographic criticism. 5. Semiotics. I. Fernández Ocampo, Anxo, editor.
II. Wolf, Michaela, editor.
P306.93.F73 2014
418'.020285–dc23
2014018010

ISBN: 978-0-415-71273-6 (hbk)
ISBN: 978-0-415-71274-3 (pbk)
ISBN: 978-1-315-74641-8 (ebk)

Typeset in Bembo
by Taylor & Francis Books

CONTENTS

LIST OF FIGURES

NOTES ON CONTRIBUTORS

Icíar Alonso-Araguás is Lecturer at the University of Salamanca, Spain. She holds an MA in Philosophy and a PhD in Translation and Interpreting. Her research interests include history of interpreting, public service interpreting and intercultural communication. She is a sworn translator and an active court and conference interpreter.

Brian James Baer is Professor of Russian and Translation Studies at Kent State University, USA. His research interests include translation pedagogy, translation history, and gender/sexuality in translation. He is founding editor of the journal *Translation and Interpreting Studies* (John Benjamins).

Jesús Baigorri-Jalón is former UN Staff Interpreter and Associate Professor at the University of Salamanca, Spain. He is director of the Alfaqueque research group. His research interests include the history of interpreting, public service interpreting, ideology and interpreting, and interpreters as fictional characters in literature and cinema.

Elizabeth Edwards is Research Professor and Director of the Photographic History Research Centre, De Montfort University, UK. Her publications include *Raw Histories* (2001), *Photographs Objects Histories* (2004), and *The Camera as Historian: Amateur Photographers and Historical Imagination 1885–1912* (2012).

Anxo Fernández-Ocampo is Senior Lecturer at the University of Vigo, Spain. His research focuses on the anthropology of translation. He is a member of the Translation and Paratranslation research group and creator of the ICOTI research collection, related to the modern history of translators and interpreters.

María Manuela Fernández-Sánchez is a Senior Lecturer in the Department of Translation and Interpreting at the University of Granada, Spain. A teacher of

Interpreting Techniques and Translation Theories at both undergraduate and graduate level, she has also taught abroad. She has published widely in the areas of translation theory and interpreting history.

Hilary Footitt is Senior Research Fellow in the Department of Modern Languages and European Studies at the University of Reading, UK. She has written widely on cultural relations in the 1944–45 period, and on languages in war and conflict situations.

Xoán Manuel Garrido-Vilariño gained a PhD in Translation and Interpreting with a dissertation on Translation and Paratranslation of Primo Levi's *Se questo è un uomo*. He is Lecturer for French and Galician at the University of Vigo, Spain.

Franziska Heimburger holds a Masters in history and is currently finishing a PhD in history on Allied language policy and practice on the Western Front of the First World War at the École des Hautes Études en Sciences Sociales in Paris.

Michael Kelly is Professor of French at the University of Southampton and a specialist in modern French culture and society, especially the history of ideas and intellectuals, and on public policy in the area of languages.

Pekka Kujamäki is Professor for German language at the University of Eastern Finland. He is head of the research project "In Search of Military Translation Cultures: Translation and Interpreting in World War II in Finland", funded by the Finnish Academy (2011–14).

Rachael Langford is Reader in French in the School of Modern Languages, Cardiff University, UK. Her research and publications focus on Francophone colonial and post-colonial cultural production, with a particular interest in visual culture.

Birgit Mersmann is Professor of Non-Western and European Art at the Jacobs University in Bremen, Germany. Research foci include image and media theory, contemporary East Asian and Western art, global art history, visual cultural translation and interrelations between script and image.

Kayoko Takeda is Professor of Translation and Interpreting Studies in the Graduate School of Intercultural Communication and the College of Intercultural Communication at Rikkyo University in Tokyo. Her research focuses on sociocultural and historical aspects of interpreting and translation.

Michaela Wolf is Associate Professor at the University of Graz, Austria. Research interests include translation sociology, translation history, and translation and visual anthropology, and interpreting in Nazi concentration camps. She is the author of *Die vielsprachige Seele Kakaniens. Übersetzen und Dolmetschen in der Habsburgermonarchie 1848 bis 1918* (2012).

PERMISSIONS

The editors want to thank the copyright holders for letting them reproduce their work:

Associated Press / Gtresonline; Bettmann / Corbis; Deutsches Historisches Museum; Finnish Defence Forces; Harry S. Truman Library; Imperial War Museums; John F. Kennedy Presidential Library and Museum; Library of Congress; Pitt Rivers Museum, University of Oxford.

Anxo Fernández-Ocampo provided the reproductions from the ICOTI research collection for this project.

While every effort has been made to trace the authors of the orphan works and the copyright holders and obtain permission, this has not been possible in all cases. Any omissions brought to our attention will be remedied in future editions.

2.1 *The Graphic*, 3 August 1872, p. 104. Illustrated by Arthur Boyd Houghton. Reproduced from the ICOTI coll. (144001).

2.2 *L'Illustration*, 11 August 1894, p. 116. Illustrated by Georges Scott. Article written by A. Bergougnan, pp. 116–17. Reproduced from the ICOTI coll. (145001).

2.3 *Harper's Weekly*, 23 May 1885, p. 336. Reproduced from the ICOTI coll. (144009).

3.1 Reproduced from the ICOTI coll. (214013).

3.2 Reproduced from the ICOTI coll. (214016).

3.3 Reproduced from the ICOTI coll. (214015).

3.4 Reproduced from the ICOTI coll. (214017).

3.5 Reproduced from the ICOTI coll. (214001).

4.1 © Pitt Rivers Museum, University of Oxford (1998.341.590.2) (reproduced with permission).

4.2 © Pitt Rivers Museum, University of Oxford (1998.341.282.2) (reproduced with permission).
4.3 © Pitt Rivers Museum, University of Oxford (1998.341.208.2) (reproduced with permission).
4.4 © Pitt Rivers Museum, University of Oxford (1998.341.229.2) (reproduced with permission).
5.1 Orphan work. Reproduced from the ICOTI coll. (304001).
5.2 Courtesy of Library of Congress (LC-B2-2768-6).
5.3 Courtesy of Library of Congress (LC-B2-961-3).
6.1 Reproduced from the ICOTI coll. (214020).
6.2 Reproduced from the ICOTI coll. (214007).
6.3 Reproduced from the ICOTI coll. (214019).
6.4 Reproduced from the ICOTI coll. (214008).
7.1 *Le Miroir*, 6 September 1914. Frontpage. Reproduced from the ICOTI coll. (141001).
7.2 Reproduced from the ICOTI coll. (211A06).
7.3 Reproduced from the ICOTI coll. (141002).
7.4 Reproduced from the ICOTI coll. (211A08).
8.1 Reproduced from the ICOTI coll. (211A04).
8.2 Reproduced from the ICOTI coll. (211A03).
8.3 Reproduced from the ICOTI coll. (211A09).
9.1 Photographer: A. Kühlewind. © Deutsches Historisches Museum, Berlin (PK 2011/246) (reproduced with permission).
9.2 Reproduced from the ICOTI coll. (211D06).
9.3 Reproduced from the ICOTI coll. (211D04).
10.1 © Imperial War Museums (HU 64880) (reproduced with permission).
10.2 Air Ministry Second World War official collection, H. Hensser. © Imperial War Museums (CA 13) (reproduced with permission).
10.3 War Office Second World War official collection, no. 9, Army film and photographic unit. © Imperial War Museums (IND 4836) (reproduced with permission).
10.4 War Office Second World War official collection, War Office official photographer. © Imperial War Museums (NAM 230) (reproduced with permission).
10.5 Reproduced with the permission of the estate of Clifford Charles Brady. © Imperial War Museums (ART LD 6487) (reproduced with permission).
11.1 Photographer: J. Autero. Courtesy of Finnish Defence Forces (SA-Kuva 50624).
11.2 Photographer: J. M. Vuorela. Courtesy of Finnish Defence Forces (SA-Kuva 156338).
11.3 Caption in the weekly journal *Suomen Kuvalehti* 37/1941. Photographer: J. Taube. Courtesy of Finnish Defence Forces (SA-Kuva 41915).
11.4 Photographer: J. Taube. Courtesy of Finnish Defence Forces (SA-Kuva 41916).

11.5 Photographer: J. Taube. Courtesy of Finnish Defence Forces (SA-Kuva 41918).
11.6 Photographer: J. Taube. Courtesy of Finnish Defence Forces (SA-Kuva 41921).
12.1 Orphan work. Reproduced from the ICOTI coll. (302D15).
12.2 Orphan work. Reproduced from the ICOTI coll. (302D07–1).
12.3 Orphan work. Reproduced from the ICOTI coll. (302D07–2).
13.1 Orphan work. Reproduced from the ICOTI coll. (302D12).
13.2 Orphan work. Reproduced from the ICOTI coll. (302A02).
14.1 © Jim Pringle / Associated Press / Gtresonline (AP451118022) (reproduced with permission).
14.2 © Abbie Rowe. White House Photographs. John F. Kennedy Presidential Library and Museum, Boston (R6501-B).
15.1 © Hildreth Donald / Imperial War Museums (MH 31945) (reproduced with permission).
15.2 © Bettmann / Corbis (BE081522) (reproduced with permission).
16.1 © U.S. Army. Courtesy of Harry S. Truman Library (63-1453-7).
17.1 © Associated Press / Gtresonline (u1959091800_001) (reproduced with permission).
17.2 © Associated Press / Gtresonline (u1951090800_001) (reproduced with permission).
17.3 © Abbie Rowe. White House Photographs. Courtesy of John F. Kennedy Presidential Library and Museum, Boston (AR6702-A).

FRAMING THE INTERPRETER

Michaela Wolf and Anxo Fernández-Ocampo

"Framing" is a tricky word. In the social sciences, it generally refers to a set of concepts and theoretical perspectives on how individuals, groups and societies organize, perceive and communicate about reality, but it can also refer to a prearranged "game". In the context of photography, the term designates a transformative effect that separates what is in the picture from what is not, creating visual and substantive relationships between the image's lines and shapes and its frame (Shore, 2007, pp. 54–56). The title of this book embraces all these aspects, aiming to cover many of the features which account for the complex task of constructing the interpreter figure in pictures. The volume will shed light on the various modes of representation which reveal the interpreter as a socially constructed category. In what form does the interpreter appear? Is the figure of the interpreter actively selected to compose the picture? How did the interpreter participate in the making of the picture? Which social agents are involved in composing and consuming pictures of interpreters?

Photography involves cultural dynamics that transcend norm-setting boundaries and introduce moments of continuous movement. Such moments can be found in crucial historical periods, and especially in times of conflict. In both past and present, this crossing of boundaries has entailed a labour of "translation" – whether between different languages and cultures or between different media and forms of representation. In contrast to translation between languages, translation in the shape of images and visual media opens up very new research perspectives in the social field of translating and interpreting, especially in view of the diversity of agents involved in the photographic process. These agents frame the pictures, write captions, and publish and consume photographs, producing and sharing representations in a way that offers access to the processes of constructing the public and private image of the subjects pictured – interpreters. In other words, photography allows us to uncover such processes, especially at the points where written texts reach their limits.

Visual media have a special role in shaping our ideas on contemporary history. Photographs, in particular, contribute significantly to the formation of visual signs that are indispensable to the interpretation and, simultaneously, construction of a specific historical moment (see Schwartz and Przyblyski, 2004). Photography transforms the visible, light-soaked world into a commodity, a manufactured or industrialized object that will enter the complex circuit of representation in the shape of studio photos, engravings, newspaper pictures, postcards or personal snapshots.[1] Against this background, photographs produced in spaces and periods of conflict are a particularly fruitful means for understanding historical processes and broadening our comprehension of the ways that particular constraints and conflict situations take shape. From the mid-nineteenth century on, photography was increasingly used by the press, and became the basis for reproductions of the image of interpreters. Due to the interpreter's physical visibility and the immediacy of the social relationships in which he or she participates, the interpreter seems to be the most accessible figure for translation scholars' study of the management of space and time. This links the interpreter with the photographer: the professional duties of both figures depend on that immediacy.

Translation and the visual: Shifting grounds of conflict

This book addresses some of the crucial processes at work in building the interpreting figure in situations of coercion and conflict. It considers photographs beyond their traditionally illustrative usage – without, however, neglecting that usage – and emphasizes their importance as a historical source for understanding interpreting activity in conflict situations. The focus is thus on the ways that the interpreter is represented and consumed through the camera lens. Because our object of study is the interpreter's body and its traces, the book's methodological approach to the "framing" of the interpreter uses tools taken primarily from visual anthropology (Collier and Collier, 1986; Edwards, 2009), sociology (Goffman, 1961; Bourdieu, 1965) and visual syntax (Kress and van Leeuwen, 2006) to analyse the imagery of the modern era of interpreting. This allows us to reconstruct the social and cultural practices underlying the act of photography and perceptions of that act across space and time. In addition, it helps reveal the anthropological and historical implications of the interactions between the multiple agents involved in the photographic process, opening up new avenues beyond the centrality of the written text.

In translation studies, the visual – both as an object of research and on a conceptual level – is only just beginning to attract detailed attention. Yet the relationship between translation and the visual is essential, if only because writing itself gives language a visual and spatial dimension and location. In recent years the discipline's traditional text-centredness, and the consequent dearth of methodological tools to tackle spatial dimensions of translation and interpreting, has been counterbalanced by various clusters of research addressing a visual agenda. These visual trends in translation studies are generally text-related, and apply intersemiotic approaches to extract a "translatable" text from the image. Interrelationships of text and image have primarily been studied through work on illustrated literature in translation. Several authors have shown that translations of picture books, for example, can change the interplay of the verbal and visual codes (González Davies and Oittinen, 2008; O'Sullivan, 2010; on comics, see Kaindl, 1999). The verbal and the visual may function in parallel, supporting each other, or they may contradict each other, with the illustrations telling a different story from a different perspective.

However, research on the visual dimensions of translation should not be limited to the verbal/visual continuum. Like engraving or painting, analogue photography and its derivatives, such as news journals and postcards, are material objects affecting the organization of social matters. This means that visual documents of translators may be studied in terms of the "material turn" (Bennett and Joyce, 2010) in human and social sciences and of approaches generally related to sensory experience (Howes and Marcoux, 2006). Various different fields of knowledge are currently moving within a visual environment and showing an interest in "translation" as a metaphor for transformation procedures – a trend that can be summarized under the heading of "cultural translation".[2] These initiatives principally come from art history (including architecture; for example Akcan, 2012) and contemporary art, in the latter case generally through visual installations (Muntadas, 2003; Bal and Morra, 2007) that give a spatial dimension to these procedures of translation – translation here understood as a metaphor for complex cultural transformations involving intellectual contacts, code shifts, urban planning or communication networks.

Another field of special relevance to cultural translation is anthropology, with its claim to describe alterity. This has prompted anthropology to become interested in translation's shaping of cultures (Pálsson 1994; Rubel and Rosman, 2003), while in translation studies efforts have been made to blend the sensibilities of the two disciplines, as Laplantine (1995), Buzelin (2004) or Sturge (2007) have pointed out. From the point of view of practice, translation and interpreting, scholars such as Bahadır (2004) have focused on the ethnographic study of interpreters, particularly in the field of community interpreting. Both translation and anthropology have made "distance" (in terms of spatial data, cultural gaps, scientific objectivity or exoticizing strategies) one of their main concerns, especially in the context of colonialism. Here photography plays a special role, given its potential to create distance out of decisions regarding angle, depth of field and composition. These strategies allow the photographer to produce a network of spatial and social relationships around the interpreter figure that did not exist before the picture was taken. Photography offers the human and social sciences researcher significant added value compared to language, and, while anthropology or translation studies struggle to work out abstract analyses of intercultural gaps, photography provides a more precise and sensitive record of such distance. It can also show details that are normally not perceived (Laplantine, 2009, p. 156), sometimes generating an "excess of meaning that exceeds the intention of the photographer" (Edwards, "Interpreting photographs: Some thoughts on method", in this volume).

Therefore, photography is one of the areas of anthropological research that overlaps with the visual approach to translation. Indeed, photographs themselves may be seen as acts of translation – a form of cultural translation that moves a set of experiences from one space of understanding to another, as visual ethnography does (see Edwards, "Anthropology's intermediary spaces", in this volume). If photographs can act as a valuable source of historical documentation for the activity of interpreters, then they can also be approached as a form of translation in their own right. The major focus of this book, however, will be the ways that interpreting emerged through photography.

Being produced and consumed by agents external to the interpreting profession, visual documents and representations of interpreters cast light on the visibility of the translator. However, they also show how interpreting overlaps with other professional and symbolic activities. Photographs may be studied from the perspective of social recognition, inasmuch as the interpreter figure is produced under social scrutiny and as photography (more specifically, the interplay of photographs and their captions) helps the label "interpreter" to acquire collective acceptance as a name of a profession; in addition, agents external to the act of photography itself, such as postcard and press publishers, journalists, readers, or photo and postcard recipients and their relatives and acquaintances, shape the interpreter's figure through specific records such as printed, typed or handwritten captions.

However, not all of the agents involved in the communicative flow of these photographs are external to translation practice and theory: some are themselves translation practitioners or scholars. Within translation studies, the history of translation has been the main research field to address the iconography of documentary and fictional representations of translators and interpreters. Researchers in this field have begun to focus on retracing specific historical "cultures of translation". It is through the prism of translation history that scholars have examined the iconography of the interpreter figure in historical contexts, often as a negotiator between different powers. For the case of the Conquest of the Americas, for instance, Icíar Alonso-Araguás and Jesús Baigorri-Jalón (2004) explore representations of the interpreter in images in contemporary chronicles. The figure of the dragoman, interpreting between the Ottoman Empire and the West, likewise has a long tradition of representation (see, for example, the work of Aykut Gürçağlar, 2004, on the visual identity of the Ottoman interpreters in Western painting), and María Manuela Fernández-Sánchez (2010) examines photos featuring interpreters in another setting marked by extreme relations of power, interpreting in the Cold War. All these examples show "interpreters at work": the interpreters are mostly depicted in real-life situations, and the research literature on them consequently focuses on describing the figures in their own right rather than in the context of their visual specificity, as would be possible from the methodological perspective of visual studies. In the context of popular culture, Michael Cronin (2009) looks at how the US film industry portrays the fictional interpreter for a mass audience. *Framing the Interpreter* is not unrelated to this "fictional turn" (Vieira, 1998) in translation studies, which places literary or cinematic fictional translators and interpreters at the heart of its research, but our focus is on the application of classic iconographical models to archive documents that stage interpreters in portraits or ordinary scenes. These documents are non-fictional representations, yet they often acquire a tinge of fiction, especially when the interpreters are staged in the image with other participants.

Shooting the interpreter

The prism we have chosen in order to observe visualizations of the interpreter is conflict. The reasons for this choice are dual. From the point of view of interpreting, war and other forms of violent conflict are a classic point of contact between peoples of different nations and ethnicities, both within multiethnic armies and in their encounters with friends or foes. Compared with travel narratives or commerce, for instance, war probably involves a wider social range of actors and contexts for such highly charged meetings. In fact, war is, to borrow Mary Louise Pratt's term (2008), among the "contact zones" most agonizingly dependent on language and translation. And from the point of view of photography, war has traditionally acted as a spur to technological advances and the popular diffusion of images more generally.

In translation studies, war and conflict have taken on special prominence in the wake of the "cultural turn", with its interest in the power differentials reflected in translation, not least as a way of identifying geopolitical and social injustices. Translation and interpreting activities clearly participate in shaping the ways that conflict unfolds (Baker, 2006, p. 2). Research in the last ten years has frequently highlighted the involvement of interpreters and translators in situations of military and ideological conflict (Simon, 2005; Salama-Carr, 2007; Baigorri-Jalón, 2010; Footitt and Kelly, 2012a, 2012b), emphasizing their roles in mediating and memorializing conflict (Stahuljak, 2000; Inghilleri, 2008; Wolf, 2013). The result has been the description of "translation cultures" (Prunč, 2008) that include a military culture of translation.

Situations of armed conflict thus offer special insights into the history of the interpreter figure, and specifically the part played by photographic representations in that history. In the modern era, the interpreter figure first appeared in engravings (see Fernández-Ocampo), often in civilian settings such as justice or diplomacy, but in the late nineteenth century photography took over the iconic representation of the interpreter. In this book, therefore, 50 institutional and personal photographs and 4 drawings from several historical periods of conflict are analysed in terms of the deployment of the figure of the interpreter in times of conflict and the interpreter's impact on the conflict situation.

The book also aims to show how the written text accompanying the photographs positions the interpreter inside a particular cultural and social system. Given its materiality, in most cases the photographical medium is complemented by a written text claiming that the figure in the picture is an interpreter; thus, the common characteristic of the pictures we have chosen is that they show *assumed* interpreters, socially designated as such. This raises several kinds of questions. Who decides to include or exclude the figure of an interpreter in the shot, and why? What are the implications of such inclusion? What borders, spaces or margins are created through the figure of the interpreter in the picture? What is the nature of the relationship between the picture and written discourse on the back of the photos? Some of the contributions prioritize the photographs' historical embedment, asking how the pictures can enhance our understanding of the interpreter's status in the particular domain studied (Heimburger; Takeda). Others explore how the interpreter portrayed in the pictures underpins existing norms: is he/she[3] shown in a traditional, secondary position or does the photo permit a deconstruction of dominant views of the interpreter's social secondariness (Mersmann)? All the chapters investigate the figure of the interpreter through the relationship between his or her representation in the picture and the interpreting function that the picture reveals. Of particular relevance is the photograph's capacity to support the text accompanying it, for example the captions[4] on postcards or explanations on the back of photographs.

Proposing the photographic medium as a tool for research in translation and interpreting studies, the book examines the image of the interpreter through a series of photo postcards, snapshots and press photos that were selected on the basis of these questions. These artefacts belong to a larger graphic heritage, unique in the social history and especially in the history of interpreters. Today, historical pictures in different media are disseminated via digital trading, whose economic rules (Vancassel, 2008, p. 371) imply the redistribution or reproduction of the images with new or recreated labels and categories inspired by original captions. This digitization gives the researching collector access to a kaleidoscope of images selected on the basis of parameters such as captions, and in many cases may open up more diverse perspectives than the collections stored in archives. Moreover, when selecting the photographs for this volume, we were able to use not only their explicit labelling, but also the presence in the visual artefacts themselves of some written reference to the "interpreter". Such references included newspaper articles related to the illustrations, printed captions on postcards, handwritten text on the back of photos, handwritten picture captions in private albums, typed captions pasted onto wire photographs, and keywords or tags chosen by stores and digital traders to display and resell the pictures. The references spring from a consensus reached between the photographer, the photo editor in the press and postcard industry and the photograph's consumer, be it a general audience or a more specific addressee. It is this textually explicit consensus which ensures that once domestic archives enter collectors' circuits, images continue to be indexed with the tag "interpreter". Doubly filtered – first by the producer or the first consumer who labelled the photo, then tagged again by the addressee or reseller – the "interpreter" function is revealed as a social category in its own right.

The book's research procedures are designed in accordance with these present-day networks of the distribution of photographic images, but in order to study and reproduce the photographs it was also necessary to have access to hard prints. For this purpose, researchers involved in the project were given access to the ICOTI archive (*Iconografía COntemporánea da Tradución e Interpretación*, contemporary iconography of translation and interpreting). This is a research collection of photographs and postcards that was launched in 2006 and is managed by Anxo Fernández-Ocampo at the University of Vigo. The collection does not cover the entire thematic spectrum of photographs, but may nevertheless be considered representative of the range of pictures with interpreter subjects produced over the past century in French, German and British photographic visual economies. ICOTI was not the sole supplier of material for *Framing the Interpreter*. The project's corpus combines the research collection with case studies from other sources: the photographic archives of institutions and museums such as the Deutsches Historisches Museum, the Finnish Defence Forces public archive, the Imperial War Museums and the Pitt Rivers Museum; public libraries such as the Library of Congress, the Harry S. Truman Library and the John F. Kennedy Presidential Library; and press agencies such as Bettmann / Corbis and Associated Press.

Whatever the source, the pictures were selected for their location within four specific conflict situations: the colonial period (prior to the 1930s), the First and Second World Wars and the Cold War period. Each period is associated with different photographic technologies and habits of image consumption, from conversion into postcards to the spread of domestic photography and the emergence of photography as part of the journalist's profession.[5]

Colonial period

The first set of papers addresses colonial interpreters photographed between the 1870s and the 1920s, a key period that coincides with the heyday of the "human zoos" (see Bancel *et al.*, 2002), the ethnographic exhibitions that displayed indigenous people brought from the colonies. Together, human zoos and the postcard industry became a highly efficient way of managing the rules of visual alterity at the turn of the century. Natives were dressed – or undressed – to make visible an appropriate distance from the colonizing society. Most of the contributions in this part of the book look at the figure of the native interpreter, whose activity has been described by research on individual cultural and linguistic domains subjected to colonial processes (see Lawrance *et al.*, 2006).

Colonial administrations typically organized interpreting services on parallel military and civilian scales. However, interpreting activity often fell outside the administrative framework completely, due to the social upheaval entailed by colonization, which resulted in the dissolution of the traditional ruling classes and the appointment of social outcasts to administrative positions once held by high-ranking officials. Given this situation, anthropological literature, administrative documents and even fiction reveal the complexity of the native interpreter's strategies for relaying the colonial scale of values while retaining the traditional linguistic mediator (on the *griot* in francophone Africa, see Bandia, 2005, p. 959). Nguyen Van Phong (1971) has shown how French writings on Vietnam tried to depict the social divide caused by the emergence of the new underclass of interpreters. In the same vein, several scholars (for example Mopoho, 2001; Tryuk, 2013) discuss the tensions between the "traditional" and colonial structures and the rise of new agents of linguistic and cultural mediation during the early stages of colonization.

Native interpreters were required to take part in all kind of missions, whether military, scientific, evangelical or literacy campaigns. Many of the pictures included in this volume show these indigenous auxiliaries, who, as part of the composition of the image, model the behaviour of the ideal colonized subject. At the same time, this ideal image is challenged by the conflict underlying relationships between colonial power and native society. This tension can be detected in the case of anthropological fieldwork as well, since cultural opposition between the observer and the observed also tends to reveal conflictual relations. Pictures staging colonial figures reflect the complex habitus of the mediating agents, including minors (Mersmann), who linked the colonial administration to the local population.

With the increasing diffusion of postcards in Europe, otherness was on display in a manner that foregrounded the power relations between the colonial photographer and the interpreter. The contributions in this section address that specific relationship between the photographer and the interpreter. Mersmann outlines a "colonial photographic translation", Fernández-Ocampo and Wolf describe how an editor of colonial postcards represents the subordinate nature of the interpreter by displacing his image onto the figure of his wife, while Langford shows how the photographer "mechanizes" the interpreter to achieve the desired image. And Edwards claims that photography in ethnographic fieldwork is itself an act of interpreting.

Shifting from portraits to scenes, these contributions are able to embrace the typology, the professional category and the emerging individual identity of the native interpreter.

First World War

If photographs ever deliver an authentic image of "reality", this is certainly not the case for photos taken in the First World War. The majority of individual photos taken by soldiers, depicting violence, adventure and virility, served propaganda purposes and as such still pervade our perceptions of the war. One might even say that soldiers photographed a different war from the more sanitized war of official photography. Photography gained momentum in a process of what could be called the industrialization of communication, and soon became a leading element in the mass media. In addition, the photographic eye fixed on different war events from written reports – the daily life of the civilian population in the war zones, including houses, and the clothing or customs of people in the hinterland (Holzer, 2007, p. 15). Many of the photographs also showed an "iconography of waste": acres of debris, churned-up fields and roads, and ruined dwellings (Jeffrey, 2008, p. 44).

During the war, governments on both sides tended to ban cameras from the battlefield and, in the case of the British Army, soldiers faced disciplinary action if caught taking pictures. With the progress of the war, however, governments understood the need for "military publicity", and a small number of professionals were appointed to act as photographers and cinematographers (Cochrane, 2004, p. x). They helped to create a propaganda machine, as can be seen in the huge volume of postcards that circulated during the war: 636,000 postcards were sent between July 1916 and June 1917 from Vienna alone (Schögler, 2011, p. 36).[6] Postcards were not only inexpensive to buy and send; they were also ubiquitously available and endlessly creative in the messages their pictures could convey. Simultaneously, technical progress stimulated the production of series of postcards based on single snapshots. This trend towards extensive diffusion also triggered competition between major photo producers and enabled the circulation of the same themes in both official pictures and soldiers' snaps.

In contrast to the mostly conflictual situations staged in other war photography, many pictures featuring interpreters in the First World War evoke the idea of alliance – they imply moments of understanding. Military interpreters working for both the Central Powers and the Entente were often soldiers or officers, with civilians recruited from the local population at a later stage of the war (Kraus, 2009, p. 790). Both kinds of interpreters handled communication among the troops on one side, and, in case of territorial occupations, with the local authorities. They were also involved in interrogations and in censoring the mail of prisoners of war. Two contributions in this section of the book focus on the French military interpreters who ensured that the military coalition ran smoothly (Heimburger; Kelly), emphasizing the social and operational realities behind the event staged on the various photographs. Another presents the intricate activities of interpreters in POW camps and their propagandistic function as displayed on widely distributed postcards (Wolf).

Second World War

While photographs in the First World War were distributed mainly on postcards and in the press, the Second World War saw a shift in consumption patterns, with the rise of the private photo album (Bopp, 2009), although the press continued to be the key medium for the distribution of photos. The light, cheap cameras of Agfa, Kodak or Voigtländer allowed recruits from all nations participating in the war to take photos during combat. The Leica model, in particular, was designed for a large public and was used widely in "war tourism" photography by German troops between 1939 and 1943. Later in the war, aerial photography for reconnaissance became increasingly important, and necessitated training courses for photographic experts. These training programmes accelerated technical innovation in the field and created specialists such as camera repairmen or laboratory technicians (Craven and Cate, 1955, p. 643).

In the Second World War, the propaganda function of photography reached a peak, and the camera became "a weapon in the hands of soldiers" (Spiegel

International Online, 2005). Private snapshots showed all sorts of topics related to the war and the soldiers' surroundings: picture postcard views, drinks parties, devastated villages and towns, the misery of local people, the laughter of young women. The touristic perspective of many of these photos is a common feature (see "Tourists in uniform" in Struk, 2011, pp. 52–62). On the other hand, a substantial number of these photos give evidence of crimes committed against the civilian population; the Wehrmacht soldiers, especially, demonstrated a certain systematic approach to portraying their own crimes. The *Wehrmachtsausstellung* (German Army Exhibition) created in 1995 used these private pictures to show the Wehrmacht's deep involvement in implementing a war of annihilation against Jews, prisoners of war and the civilian population (Heer, 1996).

Not surprisingly, interpreters played a major role throughout the war and on each side. The nature of their activity was not very different from that in the First World War, but the operations of war were much more complex, involving more nations and cultures, and technology was much more advanced, requiring more differentiated skills from the interpreters. Quite recently two projects have delivered valuable insights into the work of military interpreters in the field "proper", the battlefield. One is the "Languages at War" project, led by Hilary Footitt and Michael Kelly. In *Languages at War*, Footitt and Kelly (2012a) focus on the periods of "liberation" and "occupation" of parts of Western Europe after the landing of the Allied forces in 1944 and the shifting phenomena of "(non-)fraternization" with the civilian population (see Footitt and Kelly 2012a, 2012b; Footitt and Tobia, 2013). Another project is Pekka Kujamäki's "In Search of Military Translation Cultures", investigating interpreting in the Finnish–German Zone in Northern Finland between 1941 and 1945 (see Kujamäki in this volume). The publications resulting from these projects are based on intensive archival work and examine in detail the activities of military interpreters, their education, the institutionalization of their deployments, and the constraints under which they had to operate. Another research strand is the genre of memoirs written by the interpreters of the war's political protagonists, such as Paul Schmidt and Eugen Dollmann, who interpreted for Hitler, Léon Dostert for Eisenhower and A. H. Birse for Churchill. Much research has been dedicated to interpreting at the war crime trials, beginning with the Nuremberg Trials of 1945–46 (Gaiba, 1998; Baigorri-Jalón, 2004).

In view of the wide spectrum of deployments for interpreters in the Second World War, *Framing the Interpreter* cannot address more than a limited number of aspects. While Hilary Footitt's paper explores the visualization of the interpreter within the military institutions of war more generally, including issues of status, degrees of incorporation, and style, Xoán Manuel Garrido-Vilariño describes different levels of victimization embodied by the interpreters recruited in the occupation of Eastern Europe by choice or under coercion. Pekka Kujamäki discusses the possibility of identifying an assumed interpreter through contextual features, in a series of photos staging a high-ranking Red Army Officer taken prisoner by the Finns in the Continuation War (1941–44). Kayoko Takeda similarly highlights the features of the interpreter's visibility or invisibility, this time in the context of local interpreters serving foreign military occupiers.

Cold War

Finally, *Framing the Interpreter* addresses the political aspects of photographic composition against the background of the Cold War, a complex sequence of periods extending from 1947 to 1991. Pictures are chosen from the immediate post-war years until the 1960s, just before the new realism of television and colour film began to change public perceptions. The Nuremberg and Tokyo trials, milestones in interpreting history, created a paradigm shift in interpreting procedures at the very opening of this long period.

The Cold War photographs of interpreters were chosen on the basis of two settings: diplomatic traditions and military traditions. Photojournalism carried on its work in both, often publishing shots from the diplomatic arena and the battlefield in the same media, such as *Life* magazine, which was closely associated with press agencies and highly professionalized photographers. Telephotography played an active part in constructing breaking news, with professional photographers becoming involved in the drafting of captions and texts for the press agencies at home. Press photography, too, made the interpreter "readable": in wire photo technology, a typed and extended caption was pasted to the border of the original photograph, so that the caption became a part of the photographic image when it was transmitted via telegraph or telephone wires. The verbal thickness of these captions bears witness to the diversity of the interpreters' roles. In the military arena, this volume focuses on the early stages of the Korean War (1950–53) and the years of US combat involvement in the Vietnam War (1959–75) (Fernández-Sánchez), which coincided with the golden era of photojournalism. Most of the interpreters shown in photographs at this time were recruited from the military in the conflict zone. Some photographs of this later period of the Cold War reveal that interpreters at times violently interrogated uncooperative prisoners of war or civilians.[7]

Photojournalists gained prestige in war settings, but also in the diplomatic arena, recording the intense political activity in multilateral meetings and around supranational organizations such as the United Nations.[8] With the increasing visibility of interpreting in the media, coupled with the new scientific discourse around the Space Race (which highlighted translation-related issues, including machine translation; see Baer), collective consciousness now turned to the figure of the interpreter. He/she became the protagonist of urban legends, blamed for mistranslations that may have caused technical failures, as in the case of machine translation, or dangerous errors, as in that of Hiroshima. On the other hand, highly qualified diplomatic interpreters were also seen more positively, as hybrid characters skilled in diplomacy and language, able to improve the image of diplomats, and justified in publishing personal memoirs. Through its propagation of a particular image of the diplomatic interpreter in photographs, the press fostered a new professional élite. A useful compositional element that could explain the connection between two political leaders, the interpreter gained prominence in public reception of the news and took up a new position in history as a figure directly involved in newsworthy events (Alonso-Araguás).

The Cold War situation encouraged the establishment of teams of translators in administrative bodies and the placement of skilled interpreters on the professional market, thus paving the way for the first modern training institutions. As in the colonial period, photography was able to display the continuity – and the particular complicity – between photography and interpreting, especially since both photograph and interpreter were thought of as neutral conduits (Baer). It was against the backdrop of these parallel discourses that the Cold War period gave rise to contemporary theories on translation and interpreting.

Thoughts out of frame

Moving between photographic studios and everyday scenarios, *Framing the Interpreter* traces the image of the interpreter over 90 years, a period in which the emergence of black-and-white analogue photography marked the entry into modernity. The book shows how the interpreter figure is socially identified and culturally coded by pasting it into a scene. To some extent, the photographs reveal the degree of the interpreter's exposure to the social gaze, with aesthetic perspectives that are hardly – if at all – accessible through the conventional descriptive tools of translation studies. Our multidimensional perspective on the interpreter figure as represented in photographs strives to uncover the layers that make up the interpreter's activity in the specific situations studied, and to enhance understanding of the diverse ways in which social groups of different cultures use visual means to posit or change the professional settings of interpreting. Those settings, in turn, help to determine the position of the interpreter relative to other professional or social figures. By building explanatory contexts and developing interpretative frames, the contributions in this volume suggest that each culture, subculture or social group constructs its own representation of the interpreter figure, and therefore that the photographic image is an important component in the constitution of particular cultures of translation.

Framing the Interpreter is intended to offer common ground between documentary and fictional approaches in the field of the representations of the translator and interpreter: although it addresses documentary images, these share subjects and narrative procedures with fictional representations. Moreover, by deconstructing the mechanisms of the interpreter's instrumentalization, the book reveals that the photographs usually show not the interpreter directly, but a dramatization involving the figure of that interpreter, as one key actor in the picture's narrativity. On a more general level, this volume moves from analogue photography to contemporary visual stereotypes, such as those of digital private and institutional films, aiming to increase understanding of the power networks within which the interpreting figure operates in present-day societies in general and in present-day conflict situations in particular.[9]

Notes

1 The turn of the twentieth century saw a standardization of photographic formats and the emergence of a postcard industry enabled by mass-produced photolithographs and amateur editions.
2 For the development of the term in translation studies and cultural studies, see Buden and Nowotny, 2009, and Conway, 2012.
3 Partly due to the particular settings explored, almost all the interpreters investigated in this book are men.
4 Permanent captions on photo postcards grafted written text onto the photographic image (Bogdan and Weseloh, 2006, pp. 41–45), enhancing a narrative complexity which shaped the ways that the cards were issued, shared, classified and stored at home and in archives.
5 By the 1880s, the snapshot technique opened photography to an amateur public. The ability to take indoor photos was thus joined by a new fad for outdoor photos of subjects in motion, enabling the visual depiction of colonial societies and the experience of conflict.
6 The First World War was, of course, not the first military conflict to draw on the propagandistic effect of the postcard. Other prominent examples are the Boer War (1899–1902), the Boxer Rebellion in China (1900–01), the Russo-Japanese War (1904–05) or the Balkan wars (1912–13), where illustrations of this kind were deployed for similar purposes (Weigel *et al.*, 1983, p. 37).
7 See for example Corbis/Bettmann (2014), in which a Vietnamese interpreter helps to lay a towel over a suspect's face and then pours water on it.
8 FIT, the International Federation of Translators, was established in 1953. During the first stages of the Cold War, some important milestones of FIT history were reached, such as the "Translator's Charter" of 1963.
9 We would like to thank Kate Sturge for her invaluable input in style editing the papers in this volume, and Teresa Caneda Cabrera for her constructive and sensitive comments on a first version of this introduction.

References

Akcan, E., 2012. *Architecture in Translation: Germany, Turkey, and the Modern House*. Durham: Duke University Press.

Alonso-Araguás, I. and Baigorri-Jalón, J., 2004. "Iconography of interpreters in the conquest of the Americas." *TTR*, 17(1), pp. 129–53.

Bahadır, Ş., 2004. "Moving in-between: the interpreter as ethnographer and the interpreting-researcher as anthropologist." *Meta*, 49(4), pp. 805–21.

Baigorri-Jalón, J., 2004. *De Paris à Nuremberg: naissance de l'interprétation de conférence* [*From Paris to Nuremberg: The Birth of Conference Interpreting*]. Translated by C. Foz. Ottawa: Les Presses de l'Université d'Ottawa.

——, 2010. "Wars, languages and the role(s) of interpreters." In: *Les liaisons dangereuses: langues, traduction, interprétation* [*Dangerous Liaisons: Languages, Translation, Interpreting*]. Beyrouth: Université Saint-Joseph. pp. 173–204.

Baker, M., 2006. *Translation and Conflict: A Narrative Account*. London: Routledge.

Bal, M. and Morra, J., 2007. "Acts of translation." *Journal of Visual Culture*, 6(1), pp. 5–11.

Bancel, N., Blanchard, P., Boëtsch, G. and Deroo, E. eds., 2002. *Zoos humains: au temps des exhibitions humaines [Human Zoos: At the Time of Human Exhibitions]*. 2nd edition. Paris: La Découverte.

Bandia, P. F., 2005. "Esquisse d'une histoire de la traduction en Afrique" ["Sketch of a history of translation in Africa"]. *Meta*, 50(3), pp. 957–71.

Bennett, T. and Joyce, P. eds., 2010. *Material Powers: Cultural Studies, History and the Material Turn*. London: Routledge.

Bogdan, R. and Weseloh, T., 2006. *Real Photo Postcard Guide: The People's Photography*. Syracuse: Syracuse University Press.

Bopp, P., 2009. *Fremde im Visier: Fotoalben aus dem Zweiten Weltkrieg* [*Strangers in Sight: Photo Albums from the Second World War*]. Bielefeld: Kerber.

Bourdieu, P., 1965. *Un art moyen: essai sur les usages sociaux de la photographie* [*Photography: A Middle-brow Art*]. Paris: Minuit.

Buden, B. and Nowotny, S., 2009. "Cultural translation: an introduction to the problem." *Translation Studies*, 2(2), pp. 196–208.

Buzelin, H., 2004. "La traductologie, l'ethnographie et la production des connaissances." ["Translation studies, ethnography and the production of knowledge"]. *Meta*, 49(4), pp 729–46.

Cochrane, P., 2004. *The Western Front. 1916–1918*. Sydney: ABC Books.

Collier, J. and Collier, M., 1986. *Visual Anthropology: Photography as a Research Method*. Alburquerque: University of New Mexico Press.

Conway, K., 2012. "A conceptual and empirical approach to cultural translation." *Translation Studies*, 5(3), pp. 264–79.

Corbis/Bettmann, 2014. "Soldiers torturing Viet Cong suspect. 20 January 1968." *Corbis Images*. U1580563. www.corbisimages.com/stock-photo/rights-managed/U1580563/soldiers-torturing-viet-cong-suspect?popup=1 [accessed 7 April 2014].

Craven, W. F. and Cate, J. L., 1955. *The Army Air Forces in World War II*. Vol. 6. Chicago: University of Chicago Press.

Cronin, M., 2009. *Translation Goes to the Movies*. London: Routledge.

Edwards, E., 2009. "Thinking photography beyond the visual?" In: J. Long, A. Noble and E. Welch, eds. *Photography: Theoretical Snapshots*. London: Routledge. pp. 31–48.

Fernández-Sánchez, M. M., 2010. "Sobre el tratamiento de las fuentes en la historiografía posmoderna: archivos, fotografías y memorias de intérpretes en la Guerra Fría" ["On dealing with sources in postmodern historiography: archives, photographs and memories of interpreters of the Cold War"]. In: R. Rabadán, T. Guzmán-González, M. Fernández-López, eds. *Lengua, traducción, recepción: en honor de Julio César Santoyo* [*Language, Translation, Reception: In Honour of Julio César Santoyo*]. León: Universidad de León. pp. 231–46.

Footitt, H. and Kelly, M. eds., 2012a. *Languages at War: Policies and Practices of Language Contacts in Conflict*. Basingstoke: Palgrave Macmillan.

——, 2012b. *Languages and the Military: Alliances, Occupation and Peace Building*. Basingstoke: Palgrave Macmillan.

Footitt, H. and Tobia, S. eds., 2013. *War Talk: Foreign Languages and the British War Effort in Europe, 1940–47*. Basingstoke: Palgrave Macmillan.

Gaiba, F., 1998. *The Origins of Simultaneous Interpretation: The Nuremberg Trial*. Ottawa: University of Ottawa Press.

Goffman, E., 1961. *Encounters: Two Studies in the Sociology of Interaction*. Indianapolis: Bobbs-Merrill.

González Davies, M. and Oittinen, R. eds., 2008. *Whose Story? Translating the Verbal and the Visual in Literature for Young Readers*. Newcastle: Cambridge Scholars.

Gürçağlar, A., 2004. "Representation of Ottoman interpreters by Western painters." *Acta Orientalia Academiae Scientarum Hungaricae*, 57(2), pp. 231–52.

Heer, H. ed., 1996. *Verbrechen der Wehrmacht. Dimensionen des Vernichtungskrieges 1941–1944. Ausstellungskatalog* [*Crimes of the Wehrmacht: Dimensions of the War of Annihilation, 1941–1944. Exhibition catalogue*]. Hamburg: Hamburger Institut für Sozialforschung.

Holzer, A., 2007. *Die andere Front: Fotografie und Propaganda im Ersten Weltkrieg* [*The Other Front: Photography and Propaganda in the First World War*]. Darmstadt: Primus-Verlag.

Howes, D. and Marcoux, J.-S., 2006. "Introduction à la culture sensible" ["Introduction to sensing culture"]. *Anthropologie et Sociétés*, 30(3), pp. 7–17.

Inghilleri, M., 2008. "The ethical task of the translator in the geo-political arena: from Iraq to Guantánamo Bay." *Translation Studies* 1(2), pp. 212–23.

Jeffrey, I., 2008. *How to Read a Photograph: Understanding, Interpreting and Enjoying the Great Photographers*. London: Thames & Hudson.

Kaindl, K., 1999. "Thump, whizz, poom: a framework for the study of comics under translation." *Target*, 11(2), pp. 263–88.

Kraus, J., 2009. *Die feldgraue Uniformierung des deutschen Heeres 1907–1918* [*The Field Grey Uniforming of the German Army, 1907–1918*]. Vol. 1–2. Vienna: Verlag Militaria.

Kress, G. and van Leeuwen, T., 2006. *Reading Images: The Grammar of Visual Design*. 2nd edition. London: Routledge.

Laplantine, F., 1995. "L'ethnologue, le traducteur et l'écrivain" ["The ethnologist, the translator and the writer"]. *Meta*, 40(3), pp. 497–507.

——, 2009. *Sons, images et langage. Anthropologie esthétique et subversion* [*Sounds, Images and Language: Aesthetic Anthropology and Subversion*]. Paris: Beauchesne.

Lawrance, B. N., Osborn, E. L. and Roberts, R. L. eds., 2006. *Intermediaries, Interpreters, and Clerks: African Employees in the Making of Colonial Africa*. Madison: University of Wisconsin Press.

Mopoho, R., 2001. "Statut de l'interprète dans l'administration coloniale en Afrique francophone" ["Status of the interpreter in colonial administration of francophone Africa"]. *Meta*, 46(3), pp. 615–26.

Muntadas, A., 2003. *Muntadas. On Translation*. Barcelona: Museu d'Art Contemporani de Barcelona.

O'Sullivan, E., 2010. "More than the sum of its parts? Synergy and picturebook translation." In: E. Di Giovanni, C. Elefante and R. Pederzoli, eds. *Écrire et traduire pour les enfants: Voix, images et mots*. [*Writing and Translating for Children: Voices, Images and Texts*]. Berne: Peter Lang. pp. 133–48.

Pálsson, G., 1994. *Beyond Boundaries: Understanding, Translation and Anthropological Discourse*. Oxford: Berg Publishers.

Pratt, M. L., 2008. *Imperial Eyes: Travel Writing and Transculturation*. London: Routledge.

Prunč, E., 2008. "Zur Konstruktion von Translationskulturen" ["Constructing translation cultures"]. In: L. Schippel, ed. *Translationskultur – ein innovatives und produktives Konzept* [*Translation Culture: An Innovative and Productive Concept*]. Berlin: Frank & Timme. pp. 19–41.

Rubel, P. G. and Rosman, A. eds., 2003. *Translating Cultures: Perspectives on Translating and Anthropology*. Oxford: Berg.

Salama-Carr, M. ed., 2007. *Translating and Interpreting Conflict*. Amsterdam: Rodopi.

Schögler, L., 2011. *Das Symbol im Bild. Propagierte Geschlechterbilder auf Postkarten im Ersten Weltkrieg in Österreich-Ungarn* [*The Symbol in the Picture: Gender Images on Postcards in the First World War in Austria-Hungary*]. Unpublished diploma thesis. Graz University.

Schwartz, V. R. and Przyblyski, J. M., 2004. *The Nineteenth-century Visual Culture Reader*. New York: Routledge.

Shore, S., 2007. *The Nature of Photographs*. London: Phaidon Press.

Simon, S. ed., 2005. "Translation and social activism." Special issue of *TTR*, 18(2), pp. 9–224.

Spiegel International Online, 2005. "Images of World War II: how the camera became a weapon." www.spiegel.de/international/images-of-world-war-ii-how-the-camera-became-a-weapon-a-339728.html [accessed 7 April 2014].

Stahuljak, Z., 2000. "Violent distortions: bearing witness to the task of wartime translators." *TTR*, 13(1), pp. 37–51.

Struk, J., 2011. *Private Pictures: Soldiers' Inside View of War*. London: I.B.Tauris.

Sturge, K., 2007. *Representing Others: Translation, Ethnography and the Museum*. Manchester: St Jerome.

Tryuk, M., 2013. "L'interprète en Afrique coloniale. Intermédiaire culturel et linguistique ou traître?" ["The interpreter in colonial Africa: cultural and linguistic mediator or betrayer?"]. *Synergies Pologne*, 10, pp. 215–24.

Vancassel, P., 2008. *Les regards photographiques: dispositifs anthropologiques et processus transindividuels* [*The Photographic Gaze: Anthropological Dispositifs and Transindividual Processes*]. PhD. Université Rennes 2.

Van Phong, N., 1971. *La société vietnamienne de 1882 à 1902 d'après les écrits des auteurs français* [*The Vietnamese Society 1882–1902 in the Writings of French Authors*]. Paris: Presses Universitaires de France.

Vieira, E. R. P., 1998. "The development of a translation paideuma and poetics in Brazil. The Campos Brothers." In: M. J. Valdés and D. Kadir, eds. *Literary Cultures of Latin America: A Comparative History*. Vol. II. Oxford: Oxford University Press. pp. 141–44.

Weigel, H., Lukan, W. and Peyfuss, M. D., 1983. *Jeder Schuss ein Russ, jeder Stoss ein Franzos. Literarische und graphische Kriegspropaganda in Deutschland und Österreich 1914–1918* [*Each Shot a Russian, Each Thrust a Frenchman: Literary and Graphic War Propaganda in Germany and Austria, 1914–1918*]. Vienna: Brandstätter.

Wolf, M., 2013. "'German speakers, step forward!': surviving through interpreting in Nazi concentration camps." *Translation and Interpreting Studies*, 8(1), pp. 1–22.

PART I

Methodological approaches

1

INTERPRETING PHOTOGRAPHS

Some thoughts on method

Elizabeth Edwards

How do we approach the various and sometimes fugitive photographs of interpreters? How do we interpret interpretations of interpreters? In an attempt to unpick at least some of this nested analytic, which is possibly one of infinite regression, I want to take a critical look at some of the commonplaces of photographic analysis, but also to consider some of the ways, beyond classic models of visual interpretation, which might help us here. This chapter is not, however, a "how-to manual" – how to interpret or translate photographs. There is a massive range of methods and opinions, descriptive, prescriptive and discursive, on questions of image interpretation – semiotics, iconology, psychoanalysis, content analysis, discourse analysis, social biography, phenomenology – and many good texts analyse their advantages and disadvantages. It is not possible or necessary to summarize them here (see, for instance, Prosser, 1998; Banks, 2001; Elkins, 2008; Jordanova, 2012; Rose, 2012; Tinkler, 2013; and, for a productive overview and critique, Rose, 2013). Rather, what follows is a contemplation on the tensions, and the possibilities, of method and approach which weave their way through the chapters in this volume.

The ontological force of photographs is their insistent relationship with "that which has been". Photographs give it immediacy, the look of the real; they hold experience in a suspended time and space as the past is made present. But simultaneously they are carefully constructed social objects, made to communicate a set of experiences to a given audience at a given time. They constitute a visual discourse which is profoundly culturally and ideologically imbricated. Yet, by their very nature, photographs outlive that intentionality to perform in other contexts, in other places and in other times, translating time, space and experience. This is their interpretative challenge.

Photographs are images and objects redolent with signifiers, the carriers of cultural meaning. Drawing on Peircian semiotics and linguistic models of analysis, many approaches to photographs have sought to decode and explain the way in which they make meaning through a quasi-lexical exploration of the signifying processes of content and form, and the cultural processes that produce and sustain those meanings. However, signifying strategies of photographs are deeply embedded within a broader visual discourse. This is constituted in a mutually sustaining process, as understandings of the world are negotiated and maintained through a series of signs which reproduce and reinforce the clusters of values making up an ideological complex.

Signification is entwined with iconographical styles. Images have formal qualities; they are iconographically shaped in ways that themselves carry meanings. A detailed formalist analysis might tell us something about the appeal or cultural accessibility of an image. The frame sets up sets of relationships – here, the styles of interpreters' proximity to major political figures. Meanings are thus produced through association with a multitude of stylistic connotations, cultural markers and recognizable tropes within the image, created through framing or camera angle, style and the arrangement of subject matter. All these are constituted through, and constitutive of, cultural and ideological values; hence, for instance, the power of the stereotype. But many other processes also come into play. Degrees of formal self-consciousness – of "compositionality" (Rose, 2001 p. 19) – vary with the intentional role of the image. Thus, an advertising image has greater compositionality than a snapshot, for instance, yet both carry meanings attributed to those categories.

Signification and beyond

While photographs can communicate through a range of signifying strategies, as icons or as symbols, it is as an index that they have been most widely discussed. This is because, as I have noted, the nature of the photograph, its trace of the real, points to the prior existence of "that which was", its referent. The reality effect of photographs, as signifiers of the real, and their signifying (content) and signified (meaning) properties are, as Roland Barthes famously argued, inseparably laminated together (Barthes, 1984, p. 6). "It was there" is possibly the primal scream of photography. However, this claim has been widely problematized, with photographs positioned on the one hand as hollow simulacra, on the other as ideologically compromised. This has perhaps resulted in a distrust of photographs as historical and cultural sources. As Georges Didi-Huberman has commented, we expect both too much and too little of photographs: ask the whole truth and we will be disappointed, for photographs are messy, inexact, inadequate to the task; ask too little and we "immediately relegat[e] them to the sphere of the *simulacrum*". The net result is the same: an inattention to photographs, because they are found somehow inadequate to the evidential task (Didi-Huberman, 2008, p. 33).

However, in most images, signifiers are fluid and recodable, dependent on the contexts of presentation and apprehension. This is a position which challenges readings premised on photographic intention. Many poststructuralist critiques, most famously Roland Barthes's (1977), renegotiate the relationship between author and reader, so that meaning does not reside necessarily in origination and intention but in the performance and reception of a statement. Thus it is readers, on whom the multiplicity of possibilities is inscribed, that make meanings through the cultural knowledge or ideological discourses they bring to the reception of photographs. Meanings are "disentangled" rather than simply "deciphered" (Barthes, 1977, pp. 147–48). For instance, with their interest in interpreters, the contributors to this volume could be said to bring a particular way of disentangling a wide variety of images. They privilege the presence and work of the interpreter in images and the signifying processes in which these are enmeshed, in order to extract certain kinds of meanings. Other readers of these photographs – perhaps someone interested in Cold War politics – would not see the interpreters as the significance of the images, but premise their reading on a different set of signifiers within the photograph. Indeed, the presence of interpreters in such a reading would arguably not be "as interpreters", but as signifying compositional elements which indicate the importance of the occasion and the political and photographic protagonists and as accoutrements to power and process. Thus, if the interpreters frame the way the photograph works, they are naturalized within these workings. Often interpreters literally frame the assumed protagonists by being inscribed at the edge of the image, visually pushing against the frame. In other cases, conversely, they are centred, integral to a group, yet surrounded by the assumed protagonists, rendered visible only on the latters' terms. Whilst they may shape the photograph visually, they are not its intentional subject.

However, there is always an energy at the edge of the frame that spills out beyond the image itself in suggestive ways. Indeed, meaning is often dependent on what is *not*, in forensic terms, in the image. A detailed forensic analysis can powerfully suggest what is beyond the frame, be it in the framing and thus exclusions of the image itself, the wider contexts of cultural visibility and invisibility, or the frames of interpretation brought to a photograph.

This brings us to a key site of the recodable potential of photographs. The instability of photographic meaning is located in the random excess of photographic inscription inherent in the medium: the camera inscribes more than was intended as "the subject" of a photograph. The layers of signification and internationality can be disrupted by this excess, which is neither selective nor hierarchical, but takes the shape of forensic detail spread across the plane of the image. Excess operates at two interrelated levels, the excess in inscription itself and the excess of meaning that exceeds the intention of the photographer. Both offer potential sites of fracture in the image. Walter Benjamin famously called the first form of excess the unconscious optics that forms the substrata of photographs; Barthes gave it a viscerally subjective quality in his concept of *punctum*. This is not merely a description of forensic detail in the image. Rather, it constitutes the destabilizing potential of that random inscription to disturb the social surface of the image.

Many of the essays in this volume engage with this first formulation of excess, but they are also premised on the second. There is a sense, as I have noted, in which the interpreters discussed in this volume were themselves often outside the intention of the image. Their presence, and indeed the act of transfer which translation implies, is assumed, opaque, and only partially visible (D'hulst, 2012, p. 142), yet it is inscribed, awaiting disentanglement, in the random excess of the images. The recodability of the photographs intersects with the excess of inscription and intention, content and meaning, that opens them to historical interrogation and enables them to be brought into analytical focus here. It is worth noting that this process points to some of the problems of quantitative "content analysis" approaches to photographs, because, although such an approach can indicate overall patterns of inclusion around, for instance, gender, it does not discriminate between different weights of significance or shifting intentional entanglements. In the essays in this book the interpreters are the major significance, but at the moment of inscription they were not. Many of the actions around photographs discussed in the essays here – the use of captions, image cropping and archiving, for instance – are attempts to control the semiotic energy of images and establish preferred readings that allow the photograph to communicate in specific visual discourses within a triangulated relationship between subject, producer and viewer.

While the semiological approaches to reading images remain important, thinking about photographic meanings has expanded from linguistically derived models of the decoded "text" to more "ethnographic" approaches that consider the effects of images over a wide range of encounters and modes of apprehension. Such approaches see the classic triangulation as inflected with a series of key modalities: technology, composition and especially the social (Rose, 2001, pp. 16–17). If the photograph remains a culturally coded product of visual discourse, practices of seeing, embodied viewing and a sense of the material encounter have become increasingly important in analysis, whether of the photograph album, the newspaper image or the digital screen.

Integral to this approach is the idea that images themselves "do" something other than simply "signify" in a lexical sense – that they elicit responses. This implies an active agency of the photographs. W.J.T. Mitchell's famous question, "What do pictures want?" thus aimed to account for "the agency, motivation, autonomy, aura, fecundity or other symptoms that make pictures into 'vital signs'" (Mitchell, 2005a, p. 6).

While this links back at one level to modes and patterns of representation, such effects cannot be reduced to these. The signifying role of the photograph here is extended beyond the visual into a whole range of social practices. According to this argument, photographs are "bundled signifiers" and their meaning-making work is constituted through co-presences of the semiotic qualities of an object such as a photograph, which is part of the tangible and sensual aspect of an engagement with the world (Keane, 2005). Similarly, Mitchell has argued that there are no "visual media" as such, but that images are "braided", in that "one sensory channel or semiotic function is woven together with another more or less seamlessly" (Mitchell, 2005b, p. 262; see also Bal, 2003). Here, signifying properties of content and iconography remain important, but extend beyond the image itself to include its material apprehension and forms of reproduction. Whether a photograph is seen and used in an archive, a newspaper or an art gallery, whether it is held in the hands or viewed on a screen – all these embodied relations with the photograph shape the viewer's response and give meaning to that image (Edwards, 2009; Rose, 2009). In turn, formal changes in those material relations, from contact print to newspaper image or from analogue to digital, will shift the image's meanings anew.

In these more anthropological considerations of the work of images, emphasis is placed on their social and material relations, their institutional framings and their spheres of action in an overall visual economy (Poole, 1997). In this model, photographic meaning is not premised solely on the image itself, on the semiotic, but is patterned by the production, ownership, dissemination, collection, institutionalization. So, while a photograph might have many "affective" qualities constituted through its representational form, it is always mediated by the purposes to which it is put. How is a photograph put to work in specific social and cultural contexts? What are its perfomative qualities? Oral historians have long used a model of social efficacy that is comprised of genre, expectancy and performance (see Edwards, 2001, pp. 20–21). Success in communication depends on adherence to a culturally determined alignment of those elements.

Approaches of this kind require an active dialogue between the content of the image and the world beyond the image frame. What does the image "do" in the world? This brings us to another commonplace of photographic analysis, the idea of "context" – that photographs must, indeed can only, be understood in relation to their context. Often, however, there is little methodological thought as to how this framing context is arrived at. In fact, it constitutes another layer of interpretative significance woven around the image. All images work within a wider visual discourse, subject to a cultural patterning of what is thinkable at a given historical moment and what can be recognized as having cultural meaning as opposed to being relegated to a sea of cultural irrelevance. Such assumptions frame the way that "context" is used to explain photographs, yet "context" is not naturally constituted but itself an act of interpretation or framing used to contain and direct meaning. Given the recodable and excessive nature of the photograph, on what grounds is its saliency assessed? What is the context of a given photograph? What are the contradictory tensions between micro- and macro-analysis, between what Pinney, following Barthes, has described as *corps* and *corpus* – between micro-event and macro-meaning, between singularity as presented in the image and the generality of history and lived experience (Pinney, 2010, pp. 25–26)? From what moment of experience and understanding does the photograph come?

The philosopher of history Frank Ankersmit has argued that the gravitational pull of context can drain the object or subject of content to the extent that the thing itself, and its statement, is "left with little to say, emptied of other possible contents" (Ankersmit, 2005, p. 279). As Ulrich Baer has argued, this happens in particular when "we paste the image into a particular type of historical understanding" (Baer, 2002, p. 3), using it to demonstrate, or even illustrate, what is already known from other sources rather than admitting a narrative worked out from the traces that present themselves from the image itself. If the positivistic drag of "context" can overpower photographs, perhaps Mieke Bal's concept of framing is more useful. She has argued that "framing" is a productive successor to the positivist and containing idea of "context", and one that implies process, activity and affect (Bal, 2002, pp. 134–35). Such a position allows for a critical and dialogic understanding of the work of photographs, and grants them performative and vital qualities.

The seriality of photographs, addressed in a number of the essays, is related to ideas of "context". Seriality emerges at many points in the "biography" of photographs, at the moment of making or that of archiving, for instance. But it becomes especially pertinent as photographic technologies change, and single images cease to be necessarily the unit of thought. Seriality negotiates between the short quotation of the single image and the longer quotation of the series. Series, of course, themselves carry an intentionality in seeking to make photographs meaningful, but their active framings point to the ways in which the image is constructed. Yet many current methods for interpreting photographs are premised on the reading of single images, rather than the analysis of collective and multifaceted visual patterns.

Thinking about series of photographs rather than single images restores time and space to an extended consideration of the image. It also creates a "natural" or organic context for the single image, and thus the sense of experience that I have noted. At a practical level, seriality also complicates readings of the content of photographs, for example in the case of approaches to proxemics, kinesics or gesture. Reading body-language, for instance, held in the moment of photographic exposure, rendered as random relationships formed within the framing, is notoriously fickle. Seriality enables embodied presences to be tracked within photography, bringing time and space back to the surface. Understanding the seriality of a photograph also forcefully indicates the ways in which selection, cropping and other processes of framing and disseminating the image have operated. Why choose this image to represent the moment of interpretation? Seriality, then, can strengthen semiotic readings as well as "anthropological" readings because it lays bare the cultural processes to which images are subjected.

All images are, however, ultimately located within social relations, both in their production and in their consumption. The new approaches to photographs I have outlined draw much of their inspiration from phenomenology, whether in the embodied experience of the viewer or in relation to content. Similarly, new thinking on photography has attempted to find methods by which to read historical potential from the surface of the image itself and its social affects. But subjectivities are also traced within photographs themselves. Ariella Azoulay, for example, has written about the civil contract of photography, trying to find a new way of negotiating subjectivities in the politics of the image (Azoulay, 2008). And Baer has argued that a photograph is a trace of a moment, however banal, that "someone has lived through", of presence and experience (Baer, 2002, p. 8). In these arguments, the recodability of the image, combined with the inscriptive trace of the photograph and a detailed forensic excavation, can render an atomic structure of that experience, as the image is imbued with both the presence of the moment and the presence of experience (Edwards, forthcoming). Such approaches have much in common with Bal's concept of "close reading" which I have already noted, and with Carlo Ginzburg's famous Morellian essay on clues and traces, in which he asserts the analytical significance of the apparently unimportant and minor detail that both shapes the macro-level and inscribes individual experience (Bal, 2002, p. 10; Ginzburg, 1983, pp. 82, 86–87). These new methods are concerned with questions of agency, and attempt to destabilize, or at least complicate, dichotomizing and ultimately pacifying notions of gaze, especially as applied to photographs. Bal's use of the word "qualified" in relation to "close reading" is important here, because this is not a return to a simplistic realism. Rather, it represents an attempt to engage with the trace of the past, here the trace of the interpreter, as the result of multiple and complex interactions. It also attempts to expand methodologically what can be read from a critical engagement with the inscribed content on the surface of the image itself, an approach that does not collapse into reliance on context alone.

Epilogue

With the exception of outright and intentional falsification, there is, in the end, no right or wrong method with photographs, no right or wrong meaning, but a series of negotiations around the historical and evidential potential of the image. This requires explanatory contexts and interpretative framings through which people make sense of photographs. In seeking to understand acts of transfer, translation and, thus, interpretation, D'hulst has called for an analysis of features that are "most accessible to observation and their interrelations, taking into account the specific configurations of agents and techniques of which the transfer features form part" (2012, p. 143). Photographs offer just such an overlooked access to acts of translation, in this case the role of interpreters and their significance. The papers presented here explore what happens when photography is brought into the centre of method and the analytical arsenal. They entangle photographs productively with other kinds of data and a thinking-through of other interpretative possibilities that, to use Walter Benjamin's wonderful analogy of linguistic translation, "envelop their content like a royal robe with ample folds" (Benjamin, 1973, p. 71).

References

Ankersmit, F., 2005. *Sublime Historical Experience*. Stanford, CA: Stanford University Press.

Azoulay, A., 2008. *The Civil Contract of Photography*. Translated by R. Mazali and R. Danieli. New York: Zone Books.

Baer, U., 2002. *Spectral Evidence: The Photography of Trauma*. Cambridge, MA: The MIT Press.

Bal, M., 2002. *Travelling Concepts in the Humanities: A Rough Guide*. Toronto: University of Toronto Press.

——, 2003. "Visual essentialism and the object of visual culture." *Journal of Visual Culture*, 2(1), pp. 5–32.

Banks, M., 2001. *Visual Methods in Social Research*. London: Sage.
Barthes, R., 1977. "The death of the author." In: R. Barthes. *Image Music Text*. Translated by S. Heath. London: Flamingo. pp. 142–48.
——, 1984. *Camera Lucida: Reflections on Photography*. Translated by R. Howard. London: Fontana.
Benjamin, W., 1973. "The task of the translator." In: W. Benjamin. *Illuminations*. Translated by H. Zohn. London: Fontana. pp. 70–82.
D'hulst, L., 2012. "(Re)locating translation history: from assumed translation to assumed transfer." *Translation Studies*, 5(2), pp. 139–55.
Didi-Huberman, G., 2008. *Images in Spite of all: Four Photographs from Auschwitz*. Translated by S. Lillis. Chicago: University of Chicago Press.
Edwards, E., 2001. *Raw Histories: Photographs, Anthropology and Museums*. Oxford: Berg.
——, 2009. "Thinking photography beyond the visual?" In: J. Long, A. Noble and E. Welch, eds. *Photography: Theoretical Snapshots*. London: Routledge. pp. 31–48.
——, forthcoming. "Facing history: photography and the challenge of presence." In: H. Wolf, ed. *Aufzeigen oder Beweisen? Die Fotografie als Kulturtechnik und Medium des Wissens* [*Showing or Proving? Photography as Cultural Technique and Medium of Knowledge*]. Berlin: Akademie Verlag.
Elkins, J., ed., 2008. *Visual Literacy*. London: Routledge.
Ginzburg, C., 1983. "Clues: Morelli, Freud, and Sherlock Holmes." In: U. Eco and T. Sebeok, eds. *The Sign of Three: Dupin, Holmes, Peirce*. Bloomington: Indiana University Press. pp. 81–118.
Jordanova, L., 2012. *The Look of the Past: Visual and Material Evidence in Historical Practice*. Cambridge: Cambridge University Press.
Keane, W., 2005. "Signs are not the grab of meaning: on the social analysis of material things." In: D. Miller, ed. *Materiality*. Durham, NC: Duke University Press. pp. 182–205.
Mitchell, W.J.T., 2005a. *What Do Pictures Want? The Lives and Loves of Images*. Chicago: University of Chicago Press.
——, 2005b. "There are no visual media." *Journal of Visual Culture*, 4(2), pp. 257–66.
Pinney, C., 2010. "Coming out better." In: W. A. Gallery and F. Winterthur, eds. *Where Three Dreams Cross: 150 Years of Photography from India, Pakistan and Bangladesh*. London: Whitechapel Art Gallery/Göttingen: Steidl. pp. 25–33.
Poole, D., 1997. *Vision, Race and Modernity: A Visual Economy of the Andean Image World*. Princeton, NJ: Princeton University Press.
Prosser, J. ed., 1998. *Image-based Research: A Sourcebook for Qualitative Researchers*. London: Falmer.
Rose, G., 2001. *Visual Methodologies: An Introduction to the Interpretation of Visual Materials*. London: Sage.
——, 2009. *Doing Family Photographs: The Domestic, the Public and the Politics of Sentiment*. Farnham: Ashgate.
——, 2012. *Visual Methodologies: An Introduction to Researching with Visual Materials*. 3rd edition. London: Sage.
——, 2013. "On the relation between 'visual research methods' and contemporary visual culture." *The Sociological Review*. http://onlinelibrary.wiley.com/doi/10.1111/1467–954X.12109/pdf [accessed 7 April 2014].
Tinkler, P., 2013. *Using Photographs in Social and Historical Research*. London: Sage.

2

ENGRAVINGS OF INTERPRETERS IN THE PHOTOGRAPHIC ERA

Anxo Fernández-Ocampo

A gaze with a long history

Many innovations in the fields of photography and printing led to the convergence of the photographic method with the techniques associated more traditionally with engraving. During the second half of the nineteenth century, it was not yet possible to directly reproduce a printed version of a photo using techniques such as halftone or heliography, but photographs were habitually used as the source for woodblock prints. The relationship between the two technologies and the growing social usage of photography have historically been determining factors in the balance between written and graphic content. In fact, illustrations from this period captioned as photos were, in reality, drawings copied from photos without major modification.[1] As the speed of information transfer increased, the art of illustrating and engraving could not keep pace and was soon abandoned; the newspapers developed their own photographic aesthetic and began to construct the information for dissemination as a function of the wide availability of images (Gervais and Morel, 2007, p. 322). People engaged in a range of professional activities – including interpreters – were captured by the photographic gaze over a period of several decades during which the printing industry perfected the techniques for reproducing engravings along with typographical characters (Samuels, 2004, p. 239).

Against this background, this contribution focuses on the specific treatment given to the image of the interpreter by the illustrated media through drawings during the heyday of photography. It opens by reviewing some historical features of the illustrated gaze on the interpreter, which stood in a long iconographical tradition. According to Paul Vancassel (2008, pp. 59–61), the "non-photographic gaze" – uses not mediated by the photographic process[2] – was a vision which, through the arts or architecture, prepared the public to understand various aspects of "the visible" (p. 105). The genres defined by the gaze prior to the arrival of mechanized photography will offer a variety of perspectives on the visible uses and institutionalization of the interpreter's work.

Depicting the interpreter at work is a recurrent strategy in the iconography of audience scenes. By decomposing the translating procedure into performed gestures, the artist articulates the syntax of the scene in a range of artistic media such as painting, embroidery, tapestry, sculpture and drawing. Similar to images showing a translator under royal patronage handing over a translation to his ruler, images of interpreting serve as linkage in a transactional schema which directs the viewer's focus to the part of the scene where political exchange is happening, and frame the area of the image reserved for the principal characters. They also function as a fully articulated device to channel the dialogue and as a way of visually connecting two participating groups in the composition of the image.

A milestone in the historical study of the iconography of interpreting is the work of Icíar Alonso-Araguás and Jesús Baigorri-Jalón (2004), who described strategies of representing verbal and non-verbal communication in the illustrated documents of the Spanish colonial expansion from the mid-sixteenth to the early seventeenth century. Prior to the end of the eighteenth century, the figure of the interpreter circulated among the learned public of the European powers in the form of portraits that allowed their viewers to recognize the traits of the dominant classes, or in composed scenes in which the interpreters embodied the political and power claims staged in the image. From the seventeenth century, diplomatic activity between the European powers and the Ottoman Empire, and the opening of trade routes in Asia, gave rise to a rich iconography of the role and clothing of the interpreter. As a case in point, Catherine Boppe-Vigne (1997) discusses the evolution of the interpreter's status through the iconography of the eighteenth and nineteenth centuries by framing the convergence of the portrait genre with diplomatic settings and in media such as costume albums. Thought of in terms of Roland Barthes's (1981) *studium*, these representations typically dignify the person or the professional role that they perform; undoubtedly, the homogenized style of dress favours the group cohesiveness of the dragoman in the administrations in contact with the Ottoman Porte until the end of the eighteenth century (Séraphin-Vincent, 1997, p. 150). Aykut Gürçağlar (2004), too, addresses the outstanding saliency and visibility of Ottoman interpreters in Western pictorial compositions at that time. Into the second half of the nineteenth century, the press remained faithful to the visual status of the state interpreter by dressing him accordingly to his duties. Engravings on diplomatic missions published in several issues of French weeklies such as *L'Illustration* and *Le Monde Illustré* showed one or several interpreters, recognizable both by the captions and by their positioning in the image.[3] However, the illustrations of this period show that the promotion of the interpreter's dignity was not an absolute rule. Some postcards depicting the stereotypical dragoman dressed like a bandit, in keeping with the orientalized iconography of travel stories (see Langford in this volume), were circulating by the turn of the twentieth century.

By this time, engraving procedures based on photos had given way to ever more efficacious methods of publishing photographs directly, so that both the photographic and non-photographic gaze converged in portrayals of the interpreter. It is therefore legitimate to ask whether the transition from engravings to photography contributed to the creation of the visual image of the interpreter in the twentieth century. This question will be addressed through three specific examples. Many engravings of portraits and group portraits were printed in the nineteenth century, but the present article deliberately sets aside those static representations and focuses instead on three dynamic group scenes: a courtroom drawing taken from the available iconography in the 1870s and 1890s issues of *L'Illustration* (Figure 2.2)[4] and two significant cases of interpretation in the colonial context, from an 1872 edition of *The Graphic* (Figure 2.1) and an 1885 edition of *Harper's Weekly* (Figure 2.3).[5]

A colonial *pas de trois*

In August 1872, the British illustrated newspaper *The Graphic* described the Henry Morton Stanley expedition to search for Livingstone. The chronicle of this expedition – which had concluded some months earlier – was based on a book by Stanley (1872), and was accompanied by engravings of the drawings and photos supplied by Stanley, who chose to represent himself in the company of his servant Kalulu and his interpreter Selim (Figure 2.1).

Using the choreographic metaphor of a *pas de trois* – a dance for three people – to describe the relative positioning of the interpreter at work, Cecilia Wadensjö

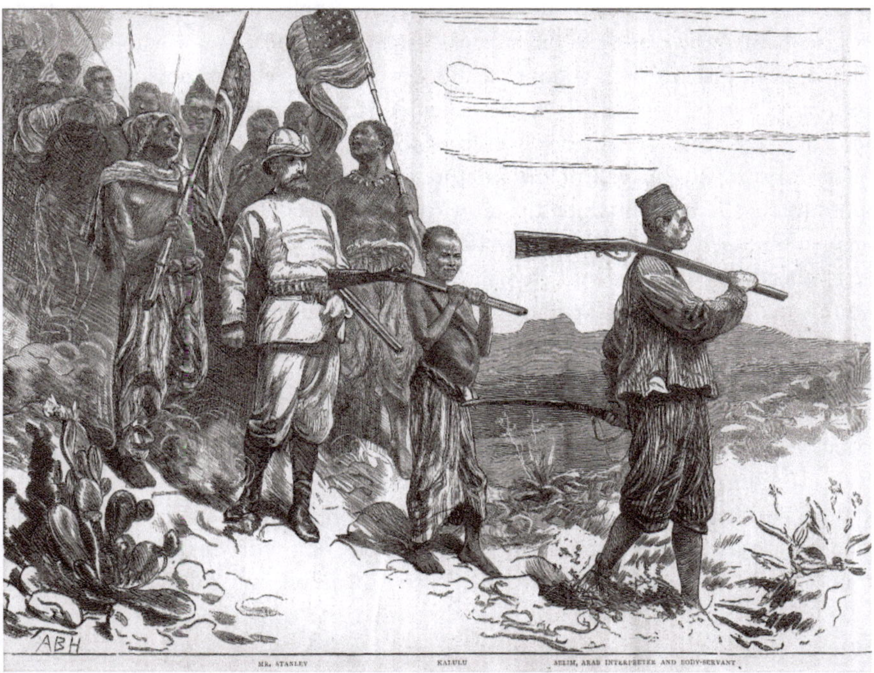

FIGURE 2.1 Stanley on his way to the coast

(1998) considered the spatial aspects of the interpretation alongside the temporal aspects of alternating speech. Similarly, Alexis Nouss (2003) compared the dance space to a translation medium, describing the dynamics of the dancer as a third space provoked by the *métissage*[6] of the body. Walking fluidly towards the edge of the illustration, the interpreter here is no longer a character in static pose. His *métis* nature is clear: Selim, neither white man nor black, is not part of the group but heads straight for the edge of the frame. The modal anthropology of François Laplantine (2005, pp. 42–44) deepens the choreographic model by focusing on the aptitude of the dancer's body to experience alterity and to expose itself to all that comes from the outside (p. 42). This model allows the space of the image to be conceived as one where transformations take place and from where the stranger and the strangeness spring.

The illustrator for *The Graphic*, Arthur Houghton, characterizes Selim the interpreter by his striped clothing and proactive position at the forefront of the colonial expedition. Selim's dynamic attitude makes this engraving into a transactional schema, fusing the scene with an event that, in cinematic terms, remains out of shot, and the narrow angle cropping the action isolates Selim at the edge of the image. He appears to be in charge of achieving the next step of the expedition. This goal is referred to in the text, creating a semantic relationship between the article and the engraving. Houghton makes the participants look older than they actually were, especially Selim, who had still been a minor when he was contracted by Stanley two years before.[7] This aesthetic ageing gives Selim a determined air, in keeping with the image of dragomans in travel writings. However, the simplicity of his clothing contrasts with the flamboyance of those dragomans. Stanley is the only figure who is standing still; it appears that Kalulu is about to take a step and that Selim is moving. Thus, the trio imitates the manner in which the expeditionary retinue starts to limber up and each character waits for the person in front to take a few steps before setting off.

The forward positioning of the interpreter is not just a whim of the artist; it reproduces a photograph which the artist copied directly. The original is captioned "Stanley with Selim and Kalulu, July 1872, Seychelles".[8] It is a studio-style portrait taken outside by an unidentified photographer. The image shows just the trio, without the rest of the expedition party, posing against a fabric backdrop to recreate a walking expedition. The name of each person appears below the image, handwritten in pencil by Stanley. It can safely be assumed that Stanley himself was responsible for staging this photo, and thus he would have been the one who placed Selim at the head of the retinue. The engraving reserves the entire right third of the image for Selim. This portion of the composition is linked to the rest by the sight lines of the rifles and Selim's sabre. In the engraving, an allegorical scene of porters appears in the background. Contrasting with this allegorical background, Selim embodies the realism of the expedition and follows the pattern suggested by Stanley, who purposely positioned the interpreter at the front to create an image of expeditionary endeavour.

The capacity to play with the impartiality of the reported story is also a central feature of another genre of drawing, the courtroom sketch.

An oblique gaze

Courtroom sketches have been a specific genre since the nineteenth century, and they occasionally include the figure of an interpreter, as in this case (Figure 2.2). The 1894 trial of the anarchist Sante Geronimo Caserio was a closely followed, high-profile event in the French press. In the following example, unlike in the remaining press coverage of the trial, *L'Illustration* opted to highlight the interpreter in both the imagery and the text. The illustration's caption ("The interrogator, the interpreter, Caserio") reveals the extent of the interpreter's mediation. In the same issue the journalist A. Bergougnan described the pictured scene and wrote about the "human" side of Caserio, quoting the interpreter Genneval to reveal Caserio's "absence of arrogance" (Bergougnan, 1894). To achieve this, Bergougnan portrayed the interpreter as being on good terms with Caserio, thus showing that Caserio was not arrogant despite the "deliberate infidelity of some of [Genneval's] translations".[9] The accompanying illustration by Georges Scott composes the scene so that the interpreter Genneval is quite prominent, with his body tilted slightly and his eyes focusing somewhere between the defendant and the jury. It is not a typical composition, as the illustrators for other publications who worked in the courtroom from a similar position to Georges Scott's omitted the interpreter in their drawings.

In this visual syntax, although there is a sight line connecting Caserio to the jury, the interpreter does not point to any other participant; moreover, with his gaze lost within the space of dialogue, the interpreter draws attention to the large empty bubble separating Caserio from the court. These in-between spaces demonstrate how an oblique view of the interpreter opens internal spaces, while stylistic and artistic practices also allow the interpreter to be dynamically linked to the dominant participants, as the engraving in Figure 2.3 shows.

To defend the rights of the Indians and "half-breeds", one of the founders of the Province of Manitoba, John Riel, rose against Canadian territorial policy and led

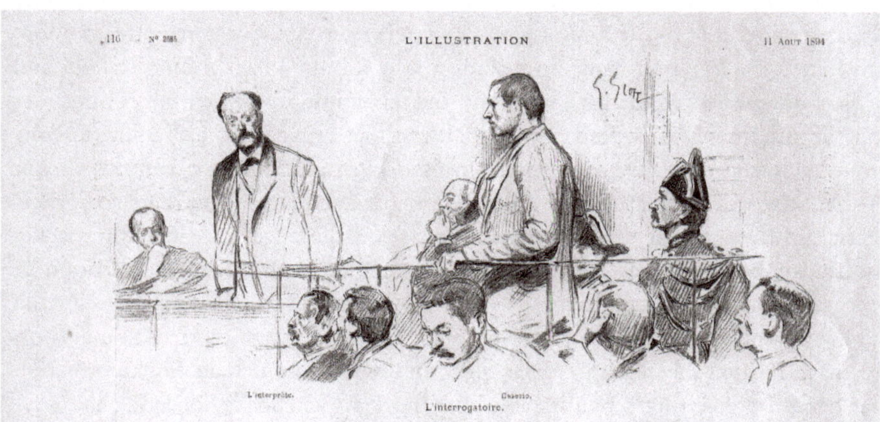

FIGURE 2.2 The Caserio trial

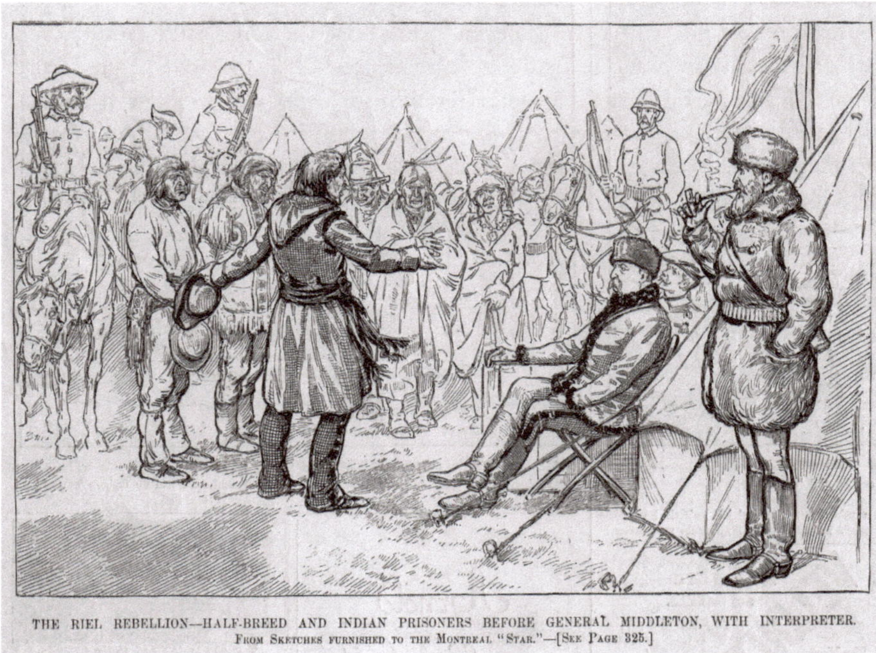

FIGURE 2.3 The Riel Rebellion. Half-breed and Indian prisoners before General Middleton, with interpreters

two rebellions in the second half of the nineteenth century. On both sides of the Canadian border, the English-language press worked to erode the legitimacy of these rebellions, among other things by giving a graphic dimension to Riel's military and political failure. On 23 May 1885, *Harper's Weekly* used an engraving to describe the surrender of a rebel group before Frederick Middleton, the officer commanding the Canadian Militia. Sitting on a folding chair, the commander directs his attention to the interpreter. The interpreter focuses the narrative tension, and addresses a fine mist made up of Indian faces against a background of tents and soldiers. Regardless of the gaze positions, the relationship between the commander and the interpreter is outlined by a bold line that emphasizes the aesthetic continuity between these two participants in the foreground. Shadow patterns chosen by the artist indicate that the very same light source converges on the two participants, and the ink that thickens the fur collar and the edge of Middleton's jacket also thickens the outlines of the standing interpreter. Line thickness, gradation of tones and composition are among the techniques with which illustrators in the publishing and journalistic spheres captured the role of the interpreter as mediator. By these means, the illustrator was able to exclude or include the information that could be given by a photograph, creating light and shade to define the relationship between the protagonists in the image – a practical impossibility for the photographic techniques of the period.

Engraving and photo print: Thickening the description of the interpreter

Through his composition of the image and the relationships between participants, as in Figure 2.3, the artist thickens the narratives on the interpreter and gives a spatial value to the mediating role. The goal of the interpreter's action may remain out of shot, as with Selim (Figure 2.1); alternatively, as evidenced by the Caserio trial illustration (Figure 2.2), the picture's drawer could cast all the participants in an artificial setting bounded by the interpreter's gaze.

In its infancy, photography could not adequately capture scenes with a group in motion, and sketches were still the most agile way of portraying current events (Gervais and Morel, 2007, p. 308). Thus, drawing was the dominant technique for courtroom sketches or visual accounts of expeditions, among other cases.[10] However, photographic equipment was better suited than drawing to producing a temporal perspective, and photography soon began to break up the illustrators' monopoly on representing the interpreter in action, especially with the arrival of photojournalism. However, the dominance of the photographic gaze since the end of the nineteenth century does not mean that the importance of engravings in understanding the history of interpreting be overlooked or forgotten; indeed, in many instances, the printed images and engravings of interpreters in nineteenth- and early-twentieth-century books and journals are the sole visual testimony to the representation of the interpreter at work. There are thus areas of research in the visual culture of interpreting that can only be analysed by taking into account the continuum between photography, illustration and engraving. This is why drawings contemporary with the development of photography allow greater understanding of the originality of the photographic gaze. From this perspective, it is possible to identify some trends that may lead to a qualitative change in our construction and use of the visual image of the interpreter in history.

The first trend concerns the mode of circulation. The turn of the twentieth century saw a growth in the exchange of photos and postcards by mail. The mechanized reproduction of the printed image and its conversion into a support for written text meant that the habitus of communication became closely linked with photography. New modes of transport and tourism, the postcard industry and the creation of private albums, among many other factors, led to scenes of conflict and colonial life – including interpreters – entering the public imagination, and the accompanying captions contributed to the widespread acceptance of the term "interpreter" for the profession.

The second trend concerns the legitimization of interpreting as a profession. It has been shown above that the non-photographic vision of the interpreter centred on engravings, paintings and drawings that evoked areas of social legitimacy connected to power and industry. Photography would reclaim this tradition, and gradually became accessible to all levels of society as the twentieth century progressed. As the cost of reproduction fell, the technical processes also became simpler and the camera became mobile. The photographer was thus able to "dedicate more attention to ways of seeing" (Vancassel, 2008, p. 369). Although human characters and material elements were occasionally part of the background of drawn and painted portraits of interpreters, as in Venetian painting, the values of the photograph, such as focus and depth of field, and the improved mobility of the camera significantly enriched the public's perception of the diversity of the interpreting profession. The snapshot conquered a further sphere of legitimacy – that of arbitrary legitimacy (Bourdieu, 1965, pp. 134–38)[11] – with the possibility of making the interpreters' social ties visible and showing how their social roles could change. Although courtroom scenes continue to be drawn by hand even today, the emergence of the instant photograph meant that photography became the main medium for portraying the professional at work.

Lastly, and as a consequence of this previous point, the third trend arises from the possibility of taking various shots of the same subject. Evidently, engraving could freely arrange the visual elements of an image in order to enhance the spatial value of the interpreter, but it was not viable to publish a whole series of engravings on the same subject. A photographer, in contrast, could shoot in series, and the resulting images could portray the interpreters in a manner more in agreement with the complexity of their role, thus compensating for the thicker narratives provided by accompanying texts of engravings (as in Figure 2.2). An early example of the possibilities of photography in extending the visual spectrum of the interpreter is the series taken by Louis Vossion in the Sudan, which shows the interpreter not as a typical individual within the racial order but in his double role as master and servant (Vossion, 1884, p. 29), either surrounded by servants or in a marginal position, half out of shot.

In both illustrations and photos, the interpreters act as a hinge in the visual syntax and introduce different temporalities. Before photography added its own vectors to the visual construction of the interpreter, such as depth of field and speed, illustrators had already discovered the value of this articulatory capacity, allowing them to compose images in which the interpreter accumulates and redistributes legitimacy.

Notes

1 In the usual procedure, a caption revealed the several stages of the printed image by naming the photographer, the drawer and the engraver. An example can be seen in *Le Monde Illustré* of 2 December 1871 (see also note 3). Further techniques led to cheaper production and printing of photos in newspapers in the early twentieth century, and the untouched photograph became the hallmark of authenticity and the basis for an impartial media.

2 Any non-photographic gaze is a form of mediation, a process for organizing reality (Vancassel, 2008, p. 61).

3 Especially from the 1860s to the 1870s, the members of Chinese and Japanese diplomatic missions to Europe and the United States were photographed, but also sketched and engraved in illustrated journals, providing a strong corpus of drawn interpreters. For instance, when the Japanese mission visiting Paris was portrayed on the cover page of *Le Monde Illustré* on 19 January 1867, one of the members figured was the interpreter

Siga-Uratarou. The picture caption on the cover page of the 2 December 1871 issue says that the Chinese ambassador Thchong-Haou and his interpreter Tshing were received by the French Prime Minister.

4 The first major publications focusing on current events through imagery were founded between 1842 and 1843: the *Illustrated London News*, the Berlin weekly *Berliner Illustrierte Zeitung* and *L'Illustration*, published in Paris.

5 All these prints are part of the ICOTI collection.

6 Nouss (2007, p. 245) explains that *métissage* should not be confused with mixture, which implies fusion, or with hybridity, which produces a new unit. Rather, it is an assemblage of affiliations and identities which is never fixed once and for all and in which the different parts retain their identity and their history.

7 The following sources from The Henry M. Stanley Archives at the Royal Museum for Central Africa in Tervuren, Belgium, were consulted to examine the contractual relationship between Selim and Stanley: Letter from Stanley to Jacob Heshmesh, 7 October 1870; Agreement between Stanley and Selim and Jacob Heshmesh, 20 January 1870; Letter of recommendation for Selim Heshmy, 18 July 1872.

8 From a print held at The Henry M. Stanley Archives (see note 7). The Seychelles photograph is not the only source for the dynamic representation of the trio, Stanley, Selim and Kalulu. *The Graphic* commissioned James Dromgole Linton to present a domesticated image of the trio for the frontispiece of the 10 August edition.

9 "En dépit de l'infidélité volontaire de quelques traductions."

10 Moreover, as Maurice Samuels (2004, p. 246) shows, silent cinema inherited from wood engraving the techniques of cutaway, insert and close-up, and in general the processes by which such engravings blended the illustrations into the text.

11 The "sphere of arbitrary legitimacy" referred to here comes from Bourdieu's theory of legitimacy as proposed in his work on photography. Bourdieu applies grades of cultural legitimacy to the arts and trades (1965, pp. 134–38) in such a manner that the sphere of the *legitimate* includes the actions of institutions and agents with the capacity to deliver symbolic capital. The sphere of the *legitimizible* covers those actions by agents, such as companies, which strive to be recognized as legitimate. The final sphere is the *arbitrary* one, in which the agents are not legitimate but develop certain actions to assert their existence before the public.

References

Alonso-Araguás, I. and Baigorri-Jalón, J., 2004. "Iconography of interpreters in the conquest of the Americas." *TTR*, 17(1), pp. 129–53.

Barthes, R., 1981. *Camera Lucida: Reflections on Photography*. Translated by R. Howard. New York: Farrar, Straus & Giroux.

Bergougnan, A., 1894. "Le procès Caserio" ["The Caserio trial"]. *L'Illustration*, 104(2685), pp. 116–17.

Boppe-Vigne, C., 1997. "L'iconographie des drogmans au XVIIIe siècle" ["The iconography of dragomans in the eighteenth century"]. In: F. Hitzel, ed. *Istanbul et les langues orientales* [*Istanbul and its Oriental Languages*]. Paris: L'Harmattan. pp. 257–70.

Bourdieu, P., 1965. *Un art moyen: essai sur les usages sociaux de la photographie* [*Photography: A Middle-brow Art*]. Paris: Minuit.

Gervais, T. and Morel, G., 2007. "Les formes de l'information. 1843–2002. De la presse illustrée aux médias modernes" ["The forms of information. 1843–2002. From the illustrated press to modern media"]. In: A. Gunthert and M. Poivert, eds. *L'Art de la photographie des origines à nos jours* [*The Art of Photography from its Origins to the Present Day*]. Paris: Citadelle et Mazenod. pp. 278–355.

Gürçağlar, A., 2004. "Representation of Ottoman interpreters by Western painters." *Acta Orientalia Academiae Scientarum Hungaricae*, 57(2), pp. 231–52.

Laplantine, F., 2005. *Le social et le sensible. Introduction à une anthropologie modale* [*The Social and the Sensitive: Introduction to a Modal Anthropology*]. Paris: Téraèdre.
Nouss, A., 2003. "Le troisième corps du break dancer" ["The third body of the break dancer"]. In: C. Fintz, ed. *Le corps comme lieu de métissages* [*The Body as a Site of Métissage*]. Paris: L'Harmattan. pp. 203–18.
——, 2007. "Translation and métissage." In: P. St-Pierre and C. K. Prafulla, eds. *Translation: Reflections, Refractions, Transformations*. Amsterdam: John Benjamins. pp. 245–52.
Samuels, M., 2004. "The illustrated history book: History between word and image." In: V. R. Schwartz and J. M. Przyblyski, eds. *The Nineteenth-century Visual Culture Reader*. London: Routledge. pp. 238–48.
Séraphin-Vincent, D., 1997. "Du drogman barataire au drogman français (1669–1793)" ["From the Ottoman drogman to the French dragoman (1669–1793)"]. In: F. Hitzel, ed. *Istanbul et les langues orientales* [*Istanbul and its Oriental Languages*]. Paris: L'Harmattan. pp. 141–52.
Stanley, H. M., 1872. *How I found Livingstone: Travels, Adventures and Discoveries in Central Africa*. New York: Scribner, Armstrong, & Co.
Vancassel, P., 2008. *Les regards photographiques: dispositifs anthropologiques et processus transindividuels* [*The Photographic Gaze: Anthropological Dispositifs and Transindividual Processes*]. Ph. D. Université Rennes 2.
Vossion, L., 1884. *Collection anthropologique du Prince Roland Bonaparte* [*Anthropological Collection of Prince Roland Bonaparte*]. Paris: Bibliothèque Nationale de France.
Wadensjö, C., 1998. *Interpreting as Interaction*. London: Longman.

PART II

Colonial exposures

3

FRAMING AND MASKING

Photographing the interpreter in/of colonial conflict

Rachael Langford

This essay considers how photographs frame and mask the role of the interpreter in colonial encounters, using Mary Louise Pratt's concept of the contact zone as a critical framework. The images considered in the essay form two groups of postcards from the European colonial period, drawn from the ICOTI research collection. Three of the photographs use a documentary-realist mode and depict the role of the colonial-era interpreter in specific situations that are alluded to by the photographs' captions or by annotations on the postcards. Two other photographs are portraits of interpreters that do not highlight the situated role of the interpreter, but reveal instead an iconography of the interpreter-guide constructed along the axis of colonial exoticism. Overall the essay will argue that this photographic corpus frames the role of the interpreter as mediator, frontier figure and dweller in – as well as delineator of – the contact zone that straddled the divide between colonial governed and colonial governing.

This notion of the contact zone is taken from Mary Louise Pratt's work and its use illuminates the photographic corpus in three ways here. First, it elucidates the historical ground from which the photographs were taken: Pratt's definition situates this ground as "the space of imperial encounters, the space in which peoples geographically and historically separated come into contact with each other and establish ongoing relations, usually involving conditions of coercion, radical inequality, and intractable conflict" (Pratt, 2008, p. 8). Pratt notes that such contact often occurs at the colonial frontier, and emphasizes that investigating imperial relations at the contact zone allows account to be taken of "co-presence, interaction, interlocking understandings and practices, and often within radically asymmetrical relations of power" (p. 8).

Second, through its applicability to the realm of representation, the notion of the contact zone provides a way of understanding the mediation of colonial reality by photographs. In this case it is the trans-temporal triangular relationship between photographer, photographed object and viewer that constitutes photographic meaning; and this meaning-creating triangle makes it possible also to consider the colonial photograph in its materiality as itself a contact zone, one that mediates changing relationships of power as the historical distance between photographer–photographed object and the viewer grows larger through time.

Third, the interpreter himself in the colonial period may be considered as embodying a contact zone: for in the relations of exchange between colonized and colonizer that Pratt outlines, the interpreter acted as a frontier figure forming and formed by the boundary between European and native, self and other. At a meta-level, it is possible to distinguish two very broad categories of official colonial interpretation roles structured by the type of colonial engagement: military interpreters incorporated into European colonial corps in situations of acute conflict during the phase of "pacification" and native auxiliaries in situations of post-"pacification" or post-conflict hegemony. Beyond these official interpreters, there lay a much more fluid and interstitial category of what could be considered interpreter–intermediary–co-creators, their own roles also structured by vectors of power, as the photographs studied here reveal. Benjamin Lawrance, *et al.* (2006, pp. 5–7) remind us that, in order to dominate in the wake of "pacification", European colonialisms relied heavily on the constant collaboration of indigenous auxiliaries such as interpreters, clerks, low-level administrators and domestic servants. Thus the necessity for interpreters and interpreter-guides in colonialism marked both the success and the impossibility of the colonial project, for colonialism's reliance on the know-how of individual native people highlighted the incoherence of colonialism as an ideology founded on a belief in backward peoples brought into enlightenment by European civilizing missions. The reality of colonialism was one of compromise, negotiation and the co-shaping of the colonial encounter, relying on the indigenous know-how of the very natives that colonialism sought to administer and dominate. Nicholas Thomas (1994, p. 16) has commented when assessing European colonialism: "it is not merely a matter of some proportion of success and failure, but also a question of exchange and the extent to which particular colonizing projects are inflected or altered by the indigenous societies they encounter". The interpreter's significant role in such exchanges, encounterings and alterations can be explored in these five colonial-era photographs.

The colonial-era interpreter photographed in situ

The first photograph for consideration (Figure 3.1) is a postcard entitled "Le Maroc – Colonne du Tadla/ 10. LES-BENI-AMIR faisant leur soumission/ L'Officier Interprète interroge leur Chef de Guerre ABDELA-BEN-JEBEUR" (Morocco – the Tadla Column/ 10. The Beni-Amir surrendering/ The interpreter officer questions their warlord Abdela Ben Jebeur). It is one of a series of postcards produced in the context of the ongoing French colonial engagements in Morocco which began around 1908. The photograph on the front of the postcard was taken by the

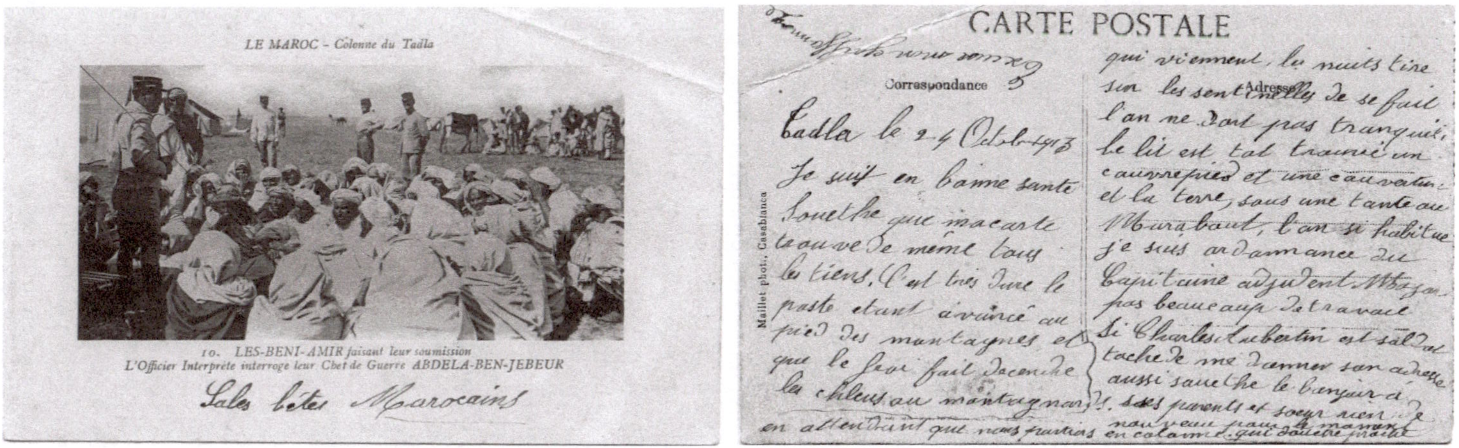

FIGURE 3.1 The Beni-Amir surrendering

Casablanca-based French photographer Maillet, who had been photographing images of architecture and life around Marrakesh since well before the establishment of the French Protectorate in 1912. The text on the back of the postcard was written in Tadla and is dated 24 October 1913. "Sales bêtes Marocains" (dirty Moroccan animals) has been handwritten below the photograph's official caption.

The revelation of this image of the colonial-era interpreter lies in its clear staging of the dynamics of contact and transaction between colonized and colonizer, and its demonstration of the "radically asymmetrical relations of power" to which Pratt refers. Here, the interpreter with his papers commands the native crowd through his singularity compared with the Moroccans' plurality; and with his upright, commanding posture compared with their crouching, submissive position. The interpreter's posture is echoed in the stance of other officers in the background, whose distribution across the frame encircles the Moroccan natives in a network of gazes that is completed by the gaze of the photographer and the viewer of the postcard: the interpreter-officer is the composition's focalizer, the point on which the concentric circles of gazes from inside the frame (that is, the near-distance officers, far-distance soldiers, warlord and Moroccans) and from outside the frame (that is, the photographer and the viewers) converge. Thus Maillet's composition draws the viewer into complicity with the hemming in and pressing together of unindividuated Moroccans, emphasizing them as a repetitive mass to be administered.

The reverse-of-postcard text was written by a private soldier from the "bataillons de tirailleurs Sénégalais", "batman" to Adjutant Captain Mbagan. This text does not refer at all to Maillet's image of an interpreter *in situ*. Instead the colonial subject writing it, himself a cog in the colonial endeavour, comments on the difficulties posed by the native enemy in a mountainous area. Maillet's photographic composition and the postcard caption place the interpreter-officer as the focal point of the image. Tellingly, however, this postcard-writing African's gaze is drawn to the lower half of the photograph as he adds the annotation "dirty Moroccan animals" to the front. The contact situation and the existence of the border-crossing figure of the interpreter passes without comment by the colonial subject, in order to emphasize the Moroccan "enemy", a distorted mirror of the self whose abject image is to be rejected.

The interpreter here is staged as a commanding figure of colonialism whose mastery of language and writing is another tool that completes the Moroccans' surrender. And yet the instability and partiality of colonial domination and of the mastery of the interpreter-officer is also inscribed in the photograph, through its opposition of the interpreter to the mass of crouched Moroccans, half-turned away from the camera and avoiding contact with the gaze of the interpreter, and through the photograph's creation of an open circle of gazes that is only closed if the direction of gaze of the viewer steadfastly replicates that of the colonial troops. This postcard therefore images the interpreter-officer as a figure who commands the gaze as well as commanding the natives through transactions of language; but the photograph also connotes the existence of contact zones where colonialism is co-shaped as an unequal transaction between agents such as the interpreter and subjects such as the Moroccans. Meanwhile, the reverse-of-postcard text demonstrates the plurality of standpoints from which colonial subjects themselves might view agents and vectors of colonial power.

The second photograph for discussion (Figure 3.2) depicts not a military interpreter fully incorporated into the structures of the colonial system, but one of the indigenous auxiliaries of colonialism studied by Lawrance *et al.* (2006). The interpreter here is also, uncomfortably, the standard-bearer for colonial rule; and his fate, as alluded to in the photograph's caption, is tied up with his intermediary role of carrying the flag, metaphorically and physically, for a particularly repressive and aggressive colonial enterprise.

The photograph dates from the end of the nineteenth century, when British colonial powers were imposing a highly unpopular Hut Tax in the British Protectorate of Sierra Leone. The caption's reference to Imperri in the Southern Province and to a later "Native Rising" (whose importance is signalled by the author's capitalization, if not by the semantics of the phrase) allows the dating of the card. Governor Cardew's decision to enforce collection of tax in 1898 contributed to widespread violent conflicts between British Protectorate forces and different African groups, culminating in a general uprising in April 1898. In the Southern Province, creole people, European and American missionaries, and African auxiliaries of colonialism were killed in large numbers in ferocious acts of resistance to British colonial government (Abraham, 1972; 1974). Against this historical context, it is clear that the interpreter and standard-bearer pictured was killed through these conflicts.

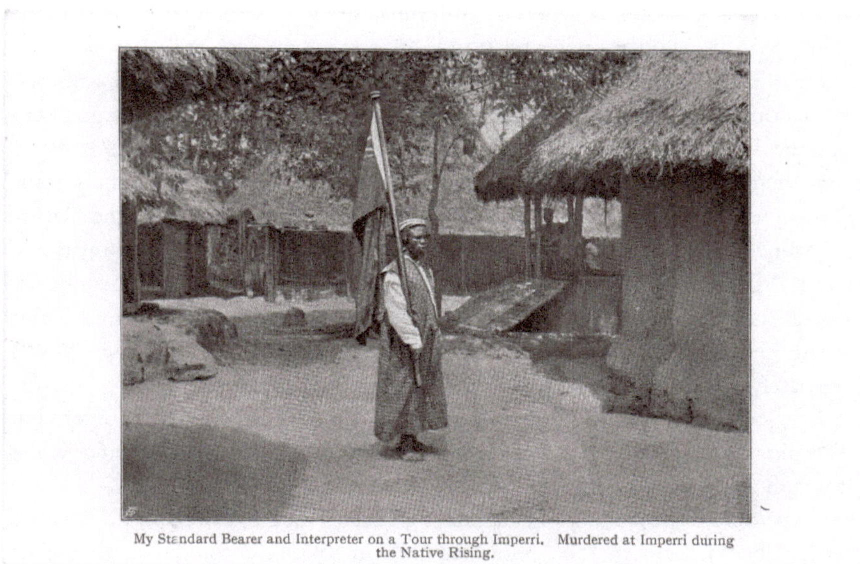

FIGURE 3.2 My standard bearer and interpreter on a tour through Imperri

This photographic representation of the interpreter–standard-bearer is a haunting image of the perils of inhabiting the contact zone. The composition emphasizes the interpreter's encirclement by indigenous dwellings while simultaneously figuring his isolation at the centre of the circle. Further, the isolated and encircled interpreter is caught between two sets of gazes: in the background of the photograph to the right, the gaze of an indigenous observer focuses directly on the interpreter; this gaze mirrors that of the photographer, positioned in front and to the right of the photographed subject. The interpreter is represented at the apex of a triangulation between colonized, interpreter–standard-bearer and colonizer–photographer; and the two sets of gazes meet on the body of the interpreter, which holds an equidistant, intermediary position between them. The gazes that meet are opposing gazes, each looking at a different side of the interpreter and metaphorizing the colonial auxiliary's isolation from both sides of the border where colonial power meets opposition to that power.

The interpreter's gaze meets neither of the gazes focused on him; he stares off into an indeterminate space between. This reinforces the interpreter as isolated and adrift, and figures him as a conduit for vectors of power rather than an embodiment of delegated power in his own right. Bearing the standard for colonialism and for the language of power "flags" this interpreter as a target for obliteration in the indigenous struggle against colonial oppression. In the photograph's caption, the possessive "my interpreter", inscribing a relationship of closeness that is highly hierarchized, jars uncomfortably with the neutral, almost dismissive tone of the note that the interpreter was "Murdered at Imperri", suggestive of the savagery of the colonized ("murdered" rather than "killed"), but also of the lack of individual emotional importance to the photographer of such an event. The image here depicts the contact zone as a perilously liminal space in which to dwell for the subaltern term in colonial power relations.

Figure 3.3 also shows the interpreter as an auxiliary of colonialism, and again sets up a relationship of possession: "l'interprète du photographe" (the photographer's interpreter) is referred to as offering tobacco leaves to the Serer women in order to prevent them from running away, and thus as facilitating the taking of the photograph. The postage stamp and knowledge of the photographic work of the Dakar-based French photographer Edmond Fortier (1862–1928) dates the image to between 1902 and 1904. It is part of the "FORTIER Photo Dakar" series consisting of several hundred numbered cards resulting from Fortier's travels around early-twentieth-century colonial Senegal (Philippe, 1986). De Smet (2009, pp. 20–21) appropriately reads this image and a cropped, coloured version showing just the interpreter's hand not as capturing elements of the real relations of power, but as recording tampered-with situations staged because of the hegemonic white photographer's presence.

This type of representation is of particular significance in the context of this essay, as is the specific function of the "photographer's interpreter"; combined, they raise questions of temporality and hierarchy (both photographer and interpreter are "translating" through different media at the same time); of multimediality (the two

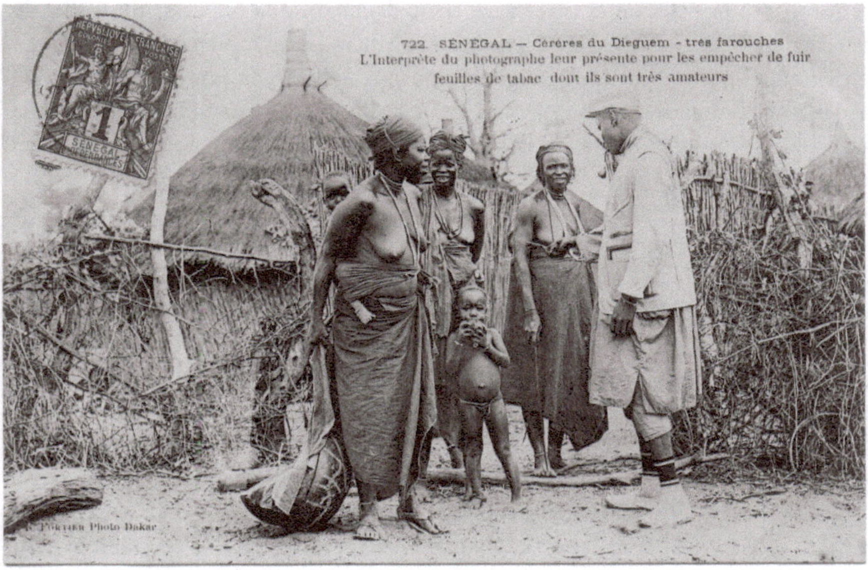

FIGURE 3.3 Serer women from Dieguem

co-produce the situation to be represented through different representational media); and of aesthetic perspective and social hierarchy. Which comes first (conceptually, practically, in cultural capital and in pragmatic hegemony), the linguistic interpretation or the photographic interpretation; and where does one end and the other begin? In this image the interpreter is not shown as a passive transmitter of cultural and linguistic knowledge, adrift between two cultures; he appears as an actant whose cultural transactions with the indigenous women make the photograph possible. The interpreter figures as a co-creator of the image, but one whose autonomy is limited by a hierarchy in which the photographer maintains rights of authorship of the image of the indigenous women and of the interpreter himself. The interpreter is therefore portrayed as the agent of the photographer; yet, paradoxically, it is this very agency that depersonalizes the interpreter into a mere mechanism for achieving the desired image.

That the interpreter is not the object of the photograph is clear from the difference between the depiction of the Serer women and that of the interpreter. The women and children are central in the image, the interpreter to one side; the women and children's facial features are evident, while the interpreter's are not visible; the women and children form a tableau of passive reception – they are not strictly engaged in any particular activity – while the interpreter extends tobacco leaves to them. Thus the image depicts the indigenous women as fixed, "being", against the interpreter's image of "doing". Moreover, the postcard's caption emphasizes the immutable nature of the identities fixed by the camera, a visual construction that is based on the opposing categories of doing and being. Thus, "l'interprète du photographe leur présente pour les empêcher de fuir feuilles de tabac" (to prevent them from running away, the photographer's interpreter offers them tobacco leaves) contrasts with the static depiction of a timeless ethnic identity expressed in "722. SÉNÉGAL – Cérères du Dieguem – très farouches" (SENEGAL – Serer women from Dieguem – very ferocious). A further contrast is set up between obscurity and enlightenment: the interpreter wears Western-influenced clothing that contrasts in the Western viewer's eyes with the indigenous women and children's lack of clothing. The interpreter's clothing is also strikingly white (in a black-and-white photograph) compared with the dark wrappers worn by the indigenous women; and in contrast to the latter's sticks and calabash, the interpreter carries in his pockets some of the tools of colonial knowledge and power, either large notebooks, or printed books, or perhaps samples of photographs.

In contrast to Figure 3.2, therefore, where the native interpreter appears fatally adrift and lost in the contact zone between two cultures, in this photograph the interpreter appears as an active bridge between past, present and future. This positive image of what an African would become in contact with colonial culture is deliberately contrasted with the image of a backward, atemporal indigenous culture fixed in the past. The photograph exemplifies Johannes Fabian's comments (1990, p. 755) that "the Other is never simply given, never just found or encountered, but made". This insight applies equally to the image of the indigenous women in the photograph as to the image of the interpreter, for the interpreter is constructed in opposition to the native and at the same time as a subordinate, active tool of the colonial photographer's eye.

Portrait of the interpreter as an "Oriental"

The second part of my corpus consists of two portraits of interpreters. These images are posed photographic confections: situations contextualizing the act of interpreting are absent and the photographs are representations of exotic, orientalist imaginings of the foreign interpreter's essence. Contextual factors informing the viewer about the interpreter's role are replaced with a close focus on indices of the interpreter's status and activity encoded not in the surroundings but in facial expression, posture, clothing and the captioning of the postcards. The interpreters portrayed in these images are not from sub-Saharan Africa, although they are depicted as non-Western natives of their respective regions. Due to colonial-era racial hierarchies that saw proximity to Caucasian physiognomy as marking gradations of humanity, these interpreters' provenance in North Africa and the Levant marks them as suitable subjects for a detailed portraiture that focuses on their individuality; the portrait is a representational mode which would have been unthinkable as a way of portraying sub-Saharan Africans. Yet the focus on the individuality of the interpreters here is not more empowering than the ways in which other Africans are portrayed, for it is based on traditions of orientalist, exoticist depictions of the non-Western foreigner. Moreover, the representation of the interpreter mediating contact and activity within a contact zone structured by unequal relations of power, which we saw in the first three photographs, recedes in Figures 3.4 and 3.5, which emphasize in contrast a static emblem of oriental

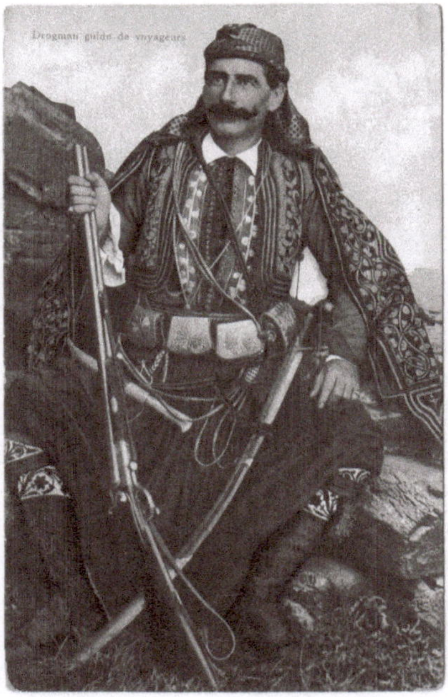

FIGURE 3.4 Dragoman, guide for travellers

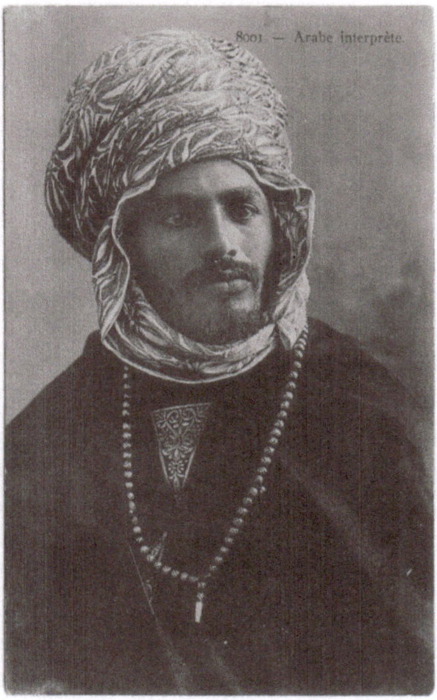

FIGURE 3.5 Arab interpreter

exoticism. In both of these portraits, the subjects do not engage directly with the camera, but look off to one side out of the image, so that the photographs allow the interpreters to be scrutinized by others but not to look back towards the viewer or photographer within the frame.

The first of the portraits (Figure 3.4) is entitled "Drogman, guide de voyageurs". It is a cropped and coloured version of an original albumen studio photograph by the prolific French photographer of the Middle East, Félix Bonfils. According to Mounira Khemir (1997, p. 199), the original photograph dates from 1865 and was taken in Lebanon, although the MIA (2014), where an original print is archived, dates the image between 1870 and 1879. The postcard's caption references the guide's role as an interpreter through the identification of the man as a "Drogman", a longstanding category of semi-official interpreters working between European and Ottoman Empire languages. In line with orientalist representational conventions of the nineteenth and early twentieth century, the subject of the portrait is the exotic (in Western eyes), heavily detailed clothing and weaponry that the dragoman displays, and not his interpreting activity. Bernard Lewis (2004, pp. 212–13) points out that the status of the dragoman in Ottoman and Western culture was subject to a number of significant changes within its long history, but that by the late nineteenth century the role had become less prominent within Ottoman government, as a growing body of students and exiles crossing from Muslim to Western countries also mastered one or more European languages. This change of status of the dragoman may be reflected to some extent in this photograph, which is not concerned with the interpreter's translational activity; indeed the caption explains the dragoman's work as that of a "travellers' guide" rather than primarily as that of a linguistic specialist. Instead, the photograph is concerned to foreground the picturesque foreign otherness of the personage, emphasizing through the seated full-body view the richly patterned clothing and the multitude of small bags and hanging accoutrements, as well as the warrior-like trappings of sword and musket which convey an alluring adventurousness to the armchair traveller and a clear differentiation between Western and Ottoman norms of dress. This sense is heightened by the paintbox-bright colouring of the archive postcard, far removed from the pale tints of the original albumen print.

What is strikingly absent from the image, however, is the notion of the interpreter inhabiting a contact zone. In this image, the zone of colonial contact becomes displaced into the photographic studio and the original relationship between photographer and dragoman; in its postcard iteration, the studio location is removed in favour of an artificial open air scene that downplays the necessity of the photographer's presence to the creation of the image, setting up a trans-temporal zone of virtual contact inhabited by both the photographed subject and the viewer of the postcard, but a zone where subject and viewer are asynchronous: any viewer may inhabit this contact zone well after the dragoman's demise simply by viewing the postcard. In this sense, where the original photograph masks the specific, located, historical context of the relations through which it came into being, the postcard contributes instead to an enduring vision of the Ottoman subject as a picturesque, exotic accessory to European travel. Thus the postcard presents a good example of the second type of contact zone discussed in the introduction, where the material object itself becomes the contact zone, a virtual space again structured by radically asymmetrical relations of power.

The final photograph of the corpus (Figure 3.5) operates similarly, although here the masking of the photographic event does not take place to the same degree, for the postcard is not as substantially altered from the photographic original as in the preceding case. The focus in this image is again on the exoticism of the subject's clothing, but the view is more derogatory than that of the preceding image. To begin with, the postcard's terse caption "Arabe interprète" ("Arab interpreter" or "interpreter Arab") gives little status or individuality to the subject: the term *Arabe*, a highly pejorative one in the French colonial context, homogenized and deindividuated what was in reality an extremely wide range of ethnicities and cultural identities. While this image's close focus on the face and head of the subject could allow the viewer to scrutinize the interpreter's features at a more individuated level, the term "Arabe" closes this possibility down by racially tagging the subject pejoratively. With clothing occupying the greater proportion of the image, the overall impression is that the interpreter's garb is the key to understanding his essence: the lower half of the frame is filled by the cloak and necklace, the upper half by the face and turban, and the headdress takes up more space in the frame as a whole than the interpreter's face. This postcard thus both instances and inflects a characteristic notion of the *Arabe* in colonial visual discourse, in which turban and cloak act as indexical features revealing more of the man who wears them than does the man's own face; and, alongside this, the postcard's caption foregrounds the subject's generic foreignness, so that the individuation granted by the subject's function – interpreter – is rescinded through the over-determined noun "Arab".

Conclusion

In this chapter, Pratt's notion of the contact zone has provided a framework for understanding the interpreter as historical agent, as cultural artefact, and as an embodiment and channeller of vectors of power. It has thus provided a privileged optic for understanding the role, status and mediations of the colonial-era interpreter. The images studied articulate representational strategies that frame and contain readings of the role of the interpreter by the use of captions, poses and stagings of the interpretational act, or present maskings of the person of the interpreter where apparel is more semantically significant than the individual. Overall, these framed and masked depictions of the colonial-era native interpreter emphasize the native interpreter's subordinate, go-between status, whether by presenting the interpreter as backward and primitive, as evolving towards European norms or as excitingly or pejoratively exotic. In contrast, the European interpreter is imaged as a paean to mastery, literacy, civilization and rightful dominance. The variety of situations and guises in which the colonial-era interpreter is figured in this small corpus reveals not only the ubiquity of the role, but also the extent to which colonial-era interpreting was defined by asymmetrical relations of power in the contact zone between dominating and dominated cultures.

References

Abraham, A., 1972. "Nyagua, the British, and the Hut Tax War." *The International Journal of African Historical Studies*, 5(1), pp. 94–98.

——, 1974. "Bai Bureh, the British, and the Hut Tax War." *The International Journal of African Historical Studies*, 7(1), pp. 99–106.

De Smet, P., 2009. *Different Truths: Ethnomedicine in Early Postcards*. Amsterdam: Kit.

Fabian, J., 1990. "Presence and representation: the other and anthropological writing." *Critical Inquiry*, 16(4), pp. 753–72.

Khemir, M., 1997. "Photography catalogue." In: R. Benjamin and M. Khemir, eds. *Orientalism Delacroix to Klee*. Sydney: Art Gallery of New South Wales. pp. 198–233.

Lawrance, B. N., Osborn, E. L. and Roberts, R. L. eds., 2006. *Intermediaries, Interpreters and Clerks: African Employees in the Making of Colonial Africa*. Madison: University of Wisconsin Press.

Lewis, B., 2004. *From Babel to Dragomans: Interpreting the Middle East*. Oxford: Oxford University Press.

MIA (Minneapolis Institute of Arts), 2014. "Drogman guide de voyageurs." https://collections.artsmia.org/index.php?page=detail&id=2959 [accessed 7 April 2014].

Philippe, D., 1986. *Inventaire général des cartes postales Fortier. 1 Séries Fortier Dakar FD et Fortier Phot. Dakar FPh* [*General Inventory of Postcards*]. Paris: Fortier.

Pratt, M. L., 2008. *Imperial Eyes: Travel Writing and Transculturation*. 2nd edition. New York: Routledge.

Thomas, N., 1994. *Colonialism's Culture: Anthropology Travel and Government*. Cambridge: Polity Press.

4

ANTHROPOLOGY'S INTERMEDIARY SPACES

Elizabeth Edwards

Anthropology is fundamentally an intercultural and interlingual act. Ideas, meanings and representations, rendered in words, are projected from one space of comprehension to another in ways that are simultaneously instrumental and hermeneutic (Rubel and Rosman, 2003, p. 6). Consequently, translation and interpretation have long been both a practice and a metaphor of anthropology. In recent years there has been extensive theoretical concern with the implications of this idea and metaphor and the problems it raises, especially as it broadens out into more expansively applied concepts of "cultural translation" (see Conway, 2012). Integral to these debates is a critical addressing of sets of assumptions, given that power relations and theoretical framings, from colonialism to gender, imbricate these positions, shaping the very relations of translation themselves (Sturge, 1997; 2007; Conway, 2012).

In the 1970s and 1980s, anthropology's interpretative processes and their politics and histories of hegemonic embeddedness – the way anthropologists "write their cultures" – was at the core of the critical destabilization and dispersal of ethnographic authority (Asad, 1986; Niranjana, 1992, pp. 47–86; Rubel and Rosman, 2003). While this moment cross-fertilized with literary studies and translation studies (Venuti, 2004, p. 325), it simultaneously drew anthropologists' actual relationships with their interlocutors into that critical view of anthropological praxis. Such debates revealed a tension between familiarization and exoticization in the act of rendering one form into another, a tension around the creation of closeness and distance (Sturge, 1997, pp. 26–33). However, while these processes have been understood and cogently argued as a series of dialogic relationships and mediations embedded in power relations, the fundamental, social act of linguistic translation itself – interpretation – remained somehow buried (Tedlock, 1983). The focus of this essay is not, then, primarily on that wider debate but on the work of photography within anthropological praxis. It aims to bring to the analytical surface the relationships on which cultural translation depended.

Anthropology's dominant method for almost a hundred years has been that of participant observation. This was, and indeed still is, premised on immediacy of experience and cultural immersion. The anthropologist lived within a community to observe from the inside while analysing from the outside. Central to this relationship, and to anthropology's own claims to authority and authenticity, has been language. Language was "the important precondition for gaining deep insights into other societies" (Borchgrevink, 2003, p. 96). As Tedlock has argued, "the anthropological dialogue creates a world, or an understanding of the *difference between* two worlds, that exists between persons who were indeterminately far apart, in all sorts of ways, when they started on their conversation" (Tedlock, 1983, p. 323).

Yet the politics and mechanisms of language acquisition and interpreting, so vital to data-gathering and the definition of evidence within the discipline, have been largely overlooked in the complex accounts of a destabilized anthropological practice, which themselves included the metaphor of "translation" itself (Jordan, 2002, p. 96). The role of interpreters in the development of anthropologists' language proficiency was key to anthropological efficacy, for "local language competence" in anthropology, on both practical and symbolic levels, was (and still is) central to its disciplinary self-positioning. It was this competence that enabled anthropologists to "understand the innermost meanings that people attach to things and events, or to discover the hidden mechanisms that make society function" (Borchgrevink, 2003, p. 96). As a result, language fills the spaces around those "outer signs of mental states" (Sturge, 1997, p. 24) upon which both ethnographic description and photography so powerfully insist. The acquisition of fluency in a largely unwritten, difficult, even obscure language was seen as both the *rite de passage* and mark of the anthropologist's authority and efficacy (Owusu, 1978, p. 312). This position has sustained the received, and indeed romantic or heroic, wisdom that anthropological fieldwork undertaken through and with interpreters is inherently flawed or second-rate. In order to maintain disciplinary mystique and authority, interpreters and linguistic mediators have thus been largely written out of the anthropological record, relegated to footnotes or brief mention in prefaces (Sturge, 2007, pp. 35–55) or described as "informants" in the translation of cultural knowledge in ways that evade the linguistic question.

Photographing Kamanga

In order to bring language, interpretation and photography into the centre of the analytical frame, I consider here a body of some 20 fieldwork photographs, taken in about 1926–27, through which interpreters and acts of cultural translation become visible. They are of a young Zande man, Kamanga, who was houseboy, interpreter, informant, clerk and friend of Edward Evans-Pritchard (hereafter EP), a British anthropologist working in south-western Sudan on the Congo border in the mid-1920s.[1]

Kamanga is nowhere described as an interpreter, but rather, as is so often the case, as an informant. However, it is clear that he had "linguistic duties" as well.

EP recalls, for instance, taking down a narrative "in Zande from his [Kamanga's] lips" and working on translations (Evans-Pritchard, 1932, p. 314). Maxwell Owusu has claimed that most African ethnography of the period was based on inadequate linguistic knowledge, with an "inevitable reliance of ethnographers on semiliterate and literate native interpreter-informants who communicated in various lingua francas" (Owusu, 1978, p. 312). In fact, despite Owusu's claim to the contrary,[2] EP was a good linguist and appears to have absorbed the Zande language.[3] In spite of extensive acknowledgements to his informants, however, there is little sense of how long this took or the role played by his interpreters. We see the interpreters only between the lines – and in the photographs.

The photographs of Kamanga from EP's fieldwork, which are now at Pitt Rivers Museum, University of Oxford, fall into two main groups: first snapshots of Kamanga, his friends and his domestic existence, and second, crucially for the consideration of translation, a series of photographs of Kamanga being initiated into the corporation of witchdoctors (*abinza*). Significantly, a sense of Kamanga's shifting and ambiguous status emerges from these two very different groups of photographs. We see EP drawn further into Kamanga's world, with its unshakable belief in spirits and the power of witchdoctors (Evans-Pritchard, 1937, p. 185), and Kamanga drawn further into EP's world of anthropological observation, colonial and intercultural relationships, and domestic arrangements.

The first, and smaller, group of photographs comprises snapshots of everyday life, for instance in the kitchen, and Kamanga with his wife, relaxing with his colleagues or occasionally with EP himself, photographs probably taken by visiting colonial officers. Overall there is little coherence within this group of images. They take their content, style, and very banality from the social practices of everyday photography. They tell not of a scientific narrative, to be returned to analytically, but of an ongoing encounter.

Particularly interesting is a portrait of Kamanga seated on a chair, placed centrally and squarely in the photographic frame (Figure 4.1). This image relates not only to photographs of EP's other informants/clerks, Mekana wiri Ongosi and Reuben wiri Rikita, but also to series of portraits of Zande elites who were photographed in this style, dressed in European clothes and often bearing the insignia of their office.[4] There is little doubt that the similarities within this series of photographs addressed not only an anthropological interest, but also Zande desires within the photographic encounter (Morton, 2012, p. 36). But Kamanga's photograph is difficult to read. It is not the production of a formal icon like the other photographs. Kamanga delicately raises his hand to his face, perhaps the gesture of a moment.

A forensic examination of the image reveals that Kamanga's left eye is puffy with a discharge – perhaps the flies were bothering him, hence the gesture? This apparently insignificant detail, a splinter of experience that reaches out from within the photograph (Baer, 2002, p. 15), reveals much about the informal relationship between EP and his informant/interpreter. The photograph is not the social icon and theatre of self of the other photographs, but may have been taken in more

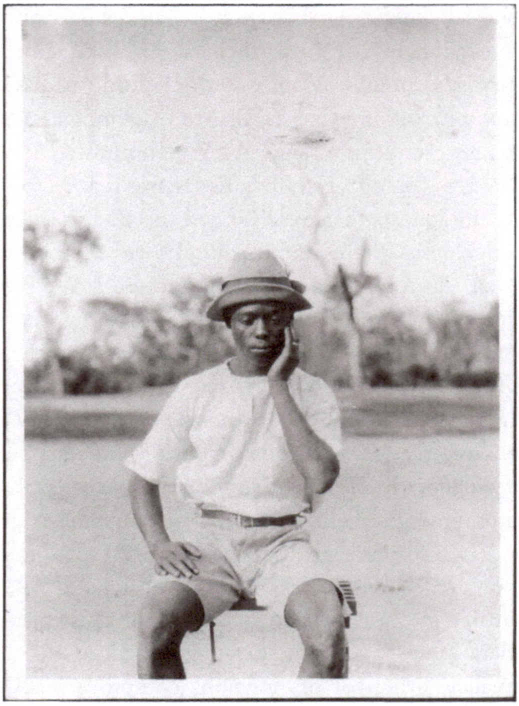

FIGURE 4.1 Kamanga

relaxed conditions. It also raises the question of the difficulty of reading gesture, held in the moment of photographic exposure. We are left only with the stilled and momentary, which is elevated to significance by the camera. This points, perhaps, to the unknowability of photographs, as "some images bypass painstaking attempts at contextualization" (Baer, 2002, p. 2) and disturb the social surface in a way similar to Barthes's visceral *punctum* (Barthes, 1984, pp. 49–53).

If Kamanga's linguistic role is ill-defined, possibly intentionally so, and the "everyday" photographs indeterminate, Kamanga is used overtly to translate culture photographically. In accordance with his practices of anthropological observation, EP's photography constitutes what might be described as "ethnographic naturalism" (Edwards, 2011, pp. 161–62). Unlike some anthropologists of the period, EP does not stage or arrange photographs for subsequent use as illustration (Morton, 2009, p. 129). Consequently, it is significant that on both of the two occasions when subjects are specifically orchestrated for the camera, Kamanga is the subject, model and mediator.

In a photograph published in EP's classic ethnography *Witchcraft, Oracles and Magic Among the Azande* (Evans-Pritchard, 1937), Kamanga is posed to demonstrate the blowing of a magic whistle. EP specifically draws attention to the act of performance and translation in this image: the plate is captioned "Kamanga Blowing a Magic Whistle (Posed)" (Figure 4.2). However, the photograph also carries visual

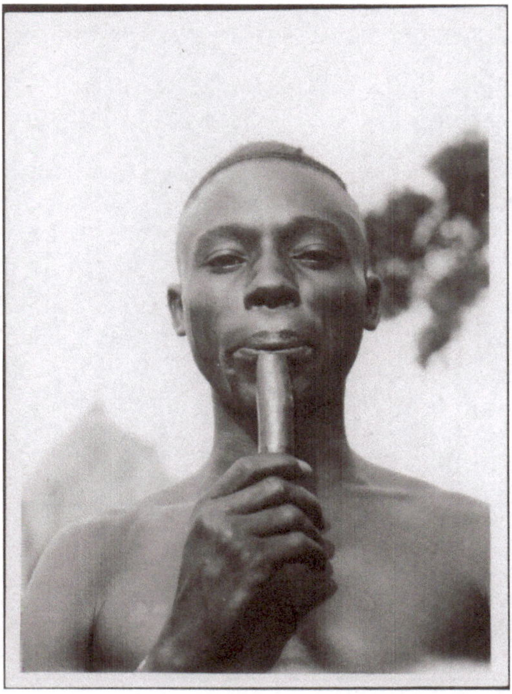

FIGURE 4.2 Kamanga blowing a magic whistle

clues to its mediated form. Even allowing for a different embouchure, Kamanga's mouth is only hesitantly and self-consciously in contact with the whistle (the low camera angle shows this clearly), and there is a directly returned engagement with the camera. Further, the close framing of the uncropped photographic negative is stylistically different from the apparently unmediated and embodied immediacy of observation which marks most of EP's field photographs (Edwards, 2011, pp. 165–66). The photograph is a self-conscious act of translation.

Of greater saliency is the other instance of performance for the camera: the photographs of Kamanga's initiation into the corporation of witchdoctors (Figure 4.3). Initiation itself is, arguably, a form of transpositional translation (Conway, 2012, p. 269) from one status to another, a transformational and transitional moment from one set of knowledge to another. Consequently, these photographs could be seen as layers of translational processes – from word and performance to image and text, attempting to transpose a subject matter which itself is a form of translation. Underlying my account of the photographs then is the slippage between different concepts of translation, from practices to metaphor, from "rewriting" to "transposition", nested within the act of anthropological observation and inscription which encompasses the oral, written, photographic, temporal, physical, social contextual, and performative (Sturge, 1997, p. 22).

The 12 photographs that form the majority of the second group make up a visual narrative both of the Zande *abinza* initiation and incorporation rituals and of

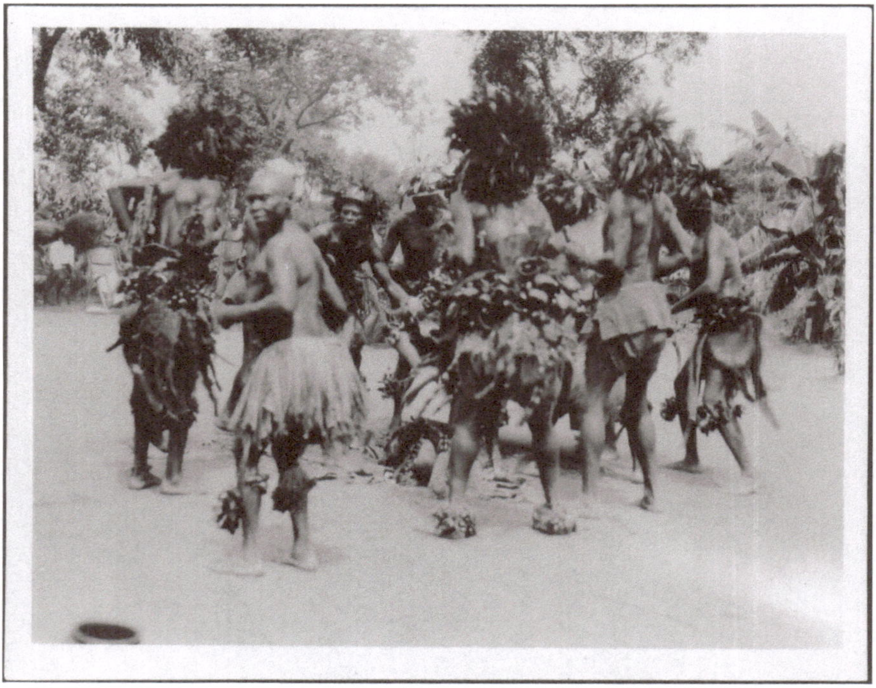

FIGURE 4.3 Witchdoctor initiation

the experiences of the anthropologist and his interpreter/informant. Christopher Morton has described and analysed the making of this sequence of photographs in detail, tracking the movement and relationships of the protagonists (Morton, 2009). Most significant for my argument are the conditions under which this photographically translated initiation occurred. This particular ceremony was sponsored by EP and took place in his compound; its *raison d'être* was for EP to take photographs. As such, it represents not only a massive extension of the "posed" photograph of Kamanga, but also a co-creation of cultural translation (Jordan, 2002, p. 98). As Morton comments, "Kamanga's lengthy training as a *binza*, and his ritual initiation into their corporation, was entirely orchestrated by [EP] in order to elicit ethnographic information" (Morton, 2005, p. 129).

EP had been unable to record properly and photograph at an earlier initiation he had witnessed. In order to acquire the photographs he needed for the later analysis and anthropological rewriting of the ritual, an initiation in which he could be closely involved had to be staged. The visual would fill the space which text had failed to fill during EP's first experience of a witchdoctor incorporation. Thus Kamanga's initiation might be regarded as an extended and self-conscious act of photographic translation, a production of ethnographic data in which the central relationship was between EP, Kamanga and the camera. Kamanga and his colleagues, such as his teacher Bögwösu, become embodiments of multiple acts of translation and interpretation imprinted on the visual narrative.

These 12 images link photography and cultural translation, as I have suggested, through the embodied presence of an informant and interpreter. In particular, photographically inscribed performance thus had a form of constitutive addressivity, defining it as an act of translation (see Bakhtin, 1986, p. 99). The photographs show the arrangement of ritual paraphernalia in the compound, the ritual burial of the initiate and the spatial dynamics of the ceremony. Throughout the narrative, EP's presence is palpable through the movement of the camera, as it gradually penetrates to the centre of the action and into the intimate relations of cultural translation. But he is not invisible: some participants turn to look at him and into his camera. The contexts of the performance are clear to all.

As if to complete the series ethnographically, there is a single, distant, still and carefully framed photograph of Kamanga which appears to have been taken at a different time, possibly later (Figure 4.4). His figure is placed centrally in the frame, "showing" cultural accoutrements and dancing dress with a self-conscious stillness that focuses attention, and perhaps resonates with the somatic mapping of earlier "types". The stillness, framing and focus of this photograph and the visual mapping of the ritual garb surely forms the basis of EP's written description (Evans-Pritchard, 1937, pp. 154–57). It positions Kamanga in his role of cultural translator at this important ethnographic moment, carrying the expectation of accurate cultural interpretation which could, in turn, shape subsequent anthropological statements.

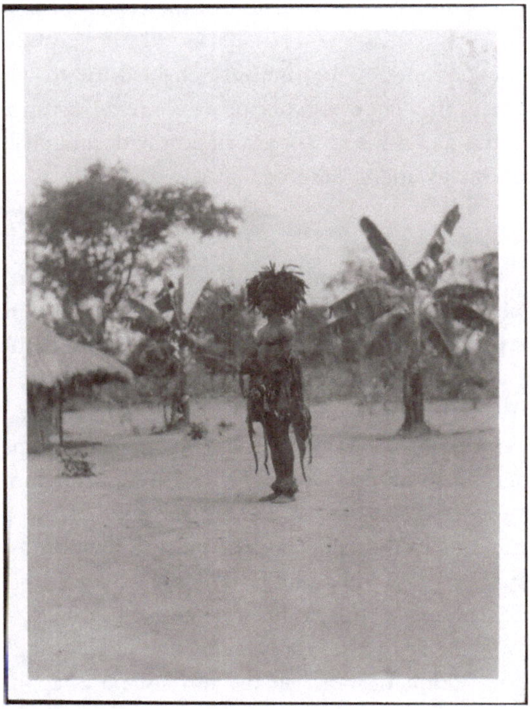

FIGURE 4.4 Kamanga in witchdoctor dancing apparel

For above all, this series of photographs can be read as visualizing the set of relationships between EP, Kamanga, the camera and Zande officiates in the initiation ritual as an act of multidimensional translation through a "myriad of small inter-linked acts of translation" (Jordan, 2002, p. 96).

One of EP's key anthropological interests in rituals such as the *abinza* initiation was their oral texts, and there is a correlation between vernacular texts and photographs as inscriptive forms of cultural practice within a process of interpretation, translation and transmediation. As Morton has noted,

> the process of writing down oral accounts, which include the problem of non-verbal communication, fragmented narratives and obscure references, seems to mirror the way in which the content of the photographs inevitably leads beyond the photographer's frame, suggesting wider contexts and alternative readings.
> *(Morton, 2009, p. 130)*

He also notes the relationship between photographs and the vernacular texts which EP collected with his interpreter-informants.[5] Texts, writing, the act of photography and photographs thus become sites where this experience is multiply inscribed: both "texts and photographs are necessarily implicated in the processes of intention, performance, translation and interpretation: both contain traces of local stories" (p. 14). Just as Kamanga becomes a site of visual translation, EP acknowledges Kamanga's role in the transcultural translation of these oral texts:

> Kamanga's sustained interest and industry enabled me to take down the gist of his experiences in a large number of native texts, given week after week for many months, and my constant association with him enabled us to discuss these texts informally and at leisure.
> *(Evans-Pritchard, 1932, p. 298)*

The integration of language, interpretation, language acquisition, cultural translation and photography is perhaps nowhere more forcefully stated than through the textual inscriptions on the back of some of EP's now-archived photographs. The simple captions are given not in the English of the archive, but in Zande (Morton, 2009, p. 134). As Walter Benjamin famously noted: "The basic error of the translator is that he preserves the state in which his own language happens to be instead of allowing his language to be powerfully affected by the foreign tongue" (1973, p. 81). These captions relate mostly to the kin relations of the people depicted. However, this is not simply the collection of kin information as part of anthropological investigation, but a "powerful and complex localizing influence over the photographs, demanding a reinsertion into the relationships surrounding their production" (Morton, 2009, p. 134). Nor is it either a "foreignizing" or "estranging" of the language or its "domestication" in the usual sense as applied to translation (Sturge, 1997, pp. 30–31; Rubel and Rosman, 2003, pp. 7–8). Rather, it represents a domestication of the photographs into Zande language as the most fitting expression of their significance and the site of their fluency as expressions of experience and presence, where the archive is best expressed by the "foreign tongue".

Closing thoughts

The photographs considered here suggest the complexity of anthropological interpretation and translation, and indicate their intersection with wider debates around the idea of "cultural translation". They blur the boundaries between inside and outside, between one form of translation and another. They also mark the many-layered experience of anthropological fieldwork. Anthropology, an essentially humanist discipline, is constituted through the human experience of complex cross-cultural encounter with a range of hopes, expectations and suspicions on both sides of the relationship.

Like translation and interpretation, photographs carry a transparency, an apparent access to an unmediated world. Yet the translator is also something of a trickster, of indeterminate allegiance and clarity, always moving across different cultural spaces to transformative effect (Rubel and Rosman, 2003, p. 16).[6] There is an ambiguity in the way in which the photograph both naturalizes through its reality effect and at the same time points to the force of the act of translation.

Photography here stands for a relationship of multiple sites of translation and interpretation which were inflected with the desires of disciplinary anthropology, the political history of the Zande region, the invisible distribution of anthropological labour and the relations of a colonialized environment (see Jezequel, 2007, p. 146). This renders the act of interpretation far from direct and simple. Interpreters operated at different levels, which are here traced photographically: "the field of knowledge production, the competing world of colonial categories, and the local arenas" (Jezequel, 2007, p. 166). Kamanga and his colleagues Mekana and Reuben were involved in complex acts of linguistic and cultural translation and co-creation of knowledge, which constituted mimetic relationships of authenticity, veracity, reality and knowledge (Niranjana, 1992, p. 50). This becomes central to the anthropological process, so that, in his initial account of the witchdoctor incorporation, EP states: "Kamanga described to me what happens in the following words … which I took down in Zande *from his lips*" (Evans-Pritchard, 1932, pp. 336, 314; emphasis added). These words, translated into English, are then quoted at length.

While I have focused on specific examples, at a conceptual level these photographs demonstrate the entanglement of photographs with multiple aspects of the broader translational process. They also act as prisms through which to think about the fundamental importance of interpreters in anthropological work, their layered presences and their fading from anthropological discourse. Yet EP made his linguistic and cultural Zande interpreters unusually visible both textually and photographically. They are scattered through his monographs, and acknowledged as collaborators in collecting and translating vernacular texts. Photography is integrally connected to EP's central field methodology of the collection of vernacular texts

(Morton, 2009, p. 120), but its contribution becomes strategically submerged and marginalized in anthropology's own historiographies, hagiographies and critiques. However, in thinking through the photographs one begins to get a sense, through a restored historicity, of the interpretative relationship as it was experienced by all concerned and how interpreters, as people, can be re-excavated.

Notes

1 Sir Edward Evan Evans-Pritchard (1902–73) was one of the most influential anthropologists of his generation. He was professor of social anthropology and fellow of All Souls College Oxford (1946–70). Throughout I am drawing on the groundbreaking work of Christopher Morton (University of Oxford), the leading authority on EP's photographs. I am very grateful to him for his comments on this paper and for sharing his detailed knowledge. Details of the original photographs discussed here can be found at the project *Recovering the Material and Visual Cultures of the Southern Sudan* (http://southernsudan.prm.ox.ac.uk/) [accessed 7 April 2014].

2 He bases his claim on EP's admission of slow and difficult language acquisition in the early stages of his Nuer fieldwork (Borchgrevink, 2003, p. 116, n. 4; Owusu, 1978, pp. 318–20).

3 While this begs the question of what constitutes linguistic competence and under what conditions (Borchgrevink, 2003, pp. 95–96), EP spoke fluent Arabic (he taught briefly at the American University in Cairo), good French and Italian, and claimed fluency in the southern Nilotic languages Nuer and Anuak as well as Zande, on which he worked hard during his first field trip of about eight months.

4 Almost all EP's assistants came from elite lineages or from families connected to local centres of power. Kamanga is described as "a courtier's son". Such people could communicate in the language of the colonial power and became important middlemen and cultural brokers (Rubel and Rosman, 2003, p. 16; Jezequel, 2007).

5 Complicating "interpreter relations", when EP left the field, he tasked Reuben wiri Rikita to continue collecting Zande texts for him. A detailed consideration of EP's textual strategies (see Sturge, 1997; 2007, pp. 47–55) is beyond the scope of this paper.

6 Significantly, the trickster is a major figure in Zande narratives. Evans-Pritchard published a number of Zande vernacular texts, collected and transcribed by his interpreters and informants, under the title *The Zande Trickster* (1967).

References

Asad, T., 1986. "The concept of cultural translation in British social anthropology." In: J. Clifford and G. Marcus, eds. *Writing Culture: The Poetics and Politics of Ethnography*. Cambridge, MA: Harvard University Press. pp. 141–64.

Baer, U., 2002. *Spectral Evidence: The Photography of Trauma*. Cambridge, MA: The MIT Press.

Bakhtin, M., 1986. *Speech Genres and Other Late Essays*.Translated by V. W. McGee. Austin: University of Texas Press.

Barthes, R., 1984. *Camera Lucida: Reflections on Photography*. Translated by R. Howard. London: Fontana.

Benjamin, W., 1973. "The task of the translator." In: W. Benjamin. *Illuminations*. Translated by H. Zohn. London: Fontana. pp. 70–82.

Borchgrevink, A., 2003. "Silencing language: of anthropologists and interpreters." *Ethnography*, 4(1), pp. 95–121.

Conway, K., 2012. "A conceptual and empirical approach to cultural translation." *Translation Studies*, 5(3), pp. 264–79.

Edwards, E., 2011. "Tracing photography." In: M. Banks and J. Ruby, eds. *Made to be Seen: Perspectives on the History of Visual Anthropology*. Chicago: University of Chicago Press. pp. 159–89.

Evans-Pritchard, E., 1932. "The Zande corporation of witchdoctors." *The Journal of the Royal Anthropological Institute of Great Britain and Ireland*, 62(2), pp. 291–336.

——, 1937. *Witchcraft, Oracles and Magic Among the Azande*. Oxford: Clarendon Press.

——. ed., 1967. *The Zande Trickster*. Oxford: Clarendon Press.

Jezequel, J.-H., 2007. "Voices of their own? African participation in the production of colonial knowledge in French West Africa, 1900–1950." In: H. Tilley and R. J. Gordon, eds. *Ordering Africa: Anthropology, European Imperialism and the Politics of Knowledge*. Manchester: Manchester University Press. pp. 145–72.

Jordan, S. A., 2002. "Ethnographic encounters: the process of cultural translation." *Language and Intercultural Communication*, 2(2), pp. 96–110.

Morton, C. 2005. "The anthropologist as photographer: reading the monograph and reading the archive." *Visual Anthropology*, 18(4), pp. 389–405.

——, 2009. "The initiation of Kamanga: visuality and textuality in Evans-Pritchard's Zande ethnography." In: C. Morton and E. Edwards, eds. *Photography, Anthropology and History: Expanding the Frame*. Farnham: Ashgate. pp. 119–42.

——, 2012. "Double alienation: Evans-Pritchard's Zande and Nuer photographs in comparative perspective." In: R. Vokes, ed. *Photography in Africa: An Ethnographic Perspective*. Woodbridge: James Currey. pp. 33–55.

Niranjana, T., 1992. *Siting Translation: History, Post-structuralism and the Colonial Context*. Berkeley: University of California Press.

Owusu, M., 1978. "Ethnography of Africa: the usefulness of the useless." *American Anthropologist*, 80(2), pp. 310–34.

Rubel, P. G. and Rosman, A. eds., 2003. *Translating Cultures: Perspectives on Translation and Anthropology*. Oxford: Berg.

Sturge, K., 1997. "Translation strategies in ethnography." *The Translator*, 3(1), pp. 21–38.

——, 2007. *Representing Others: Translation, Ethnography and the Museum*. Manchester: St Jerome.

Tedlock, D. 1983. *The Spoken Word and the Work of Interpretation*. Philadelphia: University of Pennsylvania Press.

Venuti, L. ed., 2004. *The Translation Studies Reader*. 2nd edition. London: Routledge.

5

THE MINORITIZATION OF INTERPRETING

Cultural brokerage between incapacitation and self-empowerment

Birgit Mersmann

Child interpreting has thus far played a marginal role in translation and interpreting studies. One of the main reasons for this omission might be that its agency undermines the translational paradigms, social norms and academic standards of professional interpreting – that is, professional training, linguistic and intercultural competence, knowledgeability, responsibility, social recognition and situational life experience in different communicative contexts. Since minors are not recognized as fully developed, responsible subjects and mature members of society, how can they be socially and scientifically acknowledged translators and interpreters?[1] Only in the expanding field of community interpreting and language brokering has child interpreting gained a certain attention, but it is far from bearing full fruit as an autonomous area of interpreting research.[2]

Focusing on interpreting for social and public services, community interpreting is sub-classified into legal, medical, police and asylum interpreting. The low social and often minority status of the community interpreter, most poignantly embodied by the use – or abuse – of children as interpreters, is reflected in the low value attached to community interpreting as opposed to conference interpreting and other modes or practices of interpreting. It is the lack of a strong professional identity paired with a minor socio-economic identity, and often a marginal cultural and social role, of the interpreting interlocutor that characterizes and defines community interpreting.

Besides the rather restricted meaning of community interpreting as public service interpreting in the welfare, health and legal sectors, the term "community interpreting" is more broadly identified with other forms and practices of community-related interpreting such as "community-based interpreting" (Chesher *et al.*, 2003), "escort interpreting" (Mikkelson, 1996), "liaison interpreting" (Gentile, Ozolins and Vasilakakos, 1996), "dialogue interpreting" (Mason, 2001), "ad-hoc interpreting" and "cultural interpreting" (Roberts, 1997).

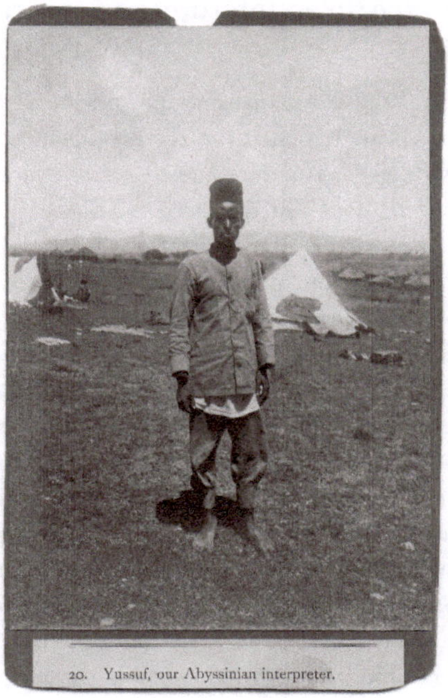

FIGURE 5.1 Yussuf, our Abyssinian interpreter

All of these broader definitional aspects of community interpreting can be related to the interpreting situation shown in the photograph of the Abyssinian interpreter Yussuf (Figure 5.1). This adolescent interpreter appears as a community-related interpreter in the sense that he serves as a representative of a particular local community (a village, settlement or tribal community) and speaks for/interprets on behalf of it. The setting is related to "escort interpreting".[3] Given this context, dialogue or ad-hoc interpreting is certainly involved in the interpreting situation. Finally, "cultural interpreting" is key to the community interpreting process at stake here, as the task of the interpreter is to cross-culturally translate "on escort" while moving directly through the field of translation.

Escort interpreting in the colonial era and migrant community interpreting by minors today share the principal situational setting: the interpreting of a minority to a majority, of an under- or even non-represented population or ethnic group to a dominant political and social group or system. The aim of this paper is to assess colonial child language brokering through the lens of ethnographic news and travel photography, as a means of visualizing interpreting history. By combining an ethnographic approach with core elements of the sociology of translation, the chapter provides a thick visual translation[4] of the interpreter as field worker. Since the historical professional situation of the adolescent escort interpreter, as captured in the photograph, is generally marked by a lack of professionalism and formalism, particular attention will be paid to the representation of his social status, identity, behaviour, manner and dress code.

The passage of the Abyssinian interpreter: Brokering cultural sovereignty

The photograph of Yussuf, the Abyssinian interpreter, shows the interpreter in the field of translation near a village settlement in the Ethiopian grasslands. He is clearly represented as a "field worker" and native interpreter. His head covered with a fez protrudes out of the picture, as it just surpasses the line of the horizon and is set against the bright sky zone that fills the upper third of the photograph. The light contrast together with the lifted-out position of the head accentuates the darkness of the interpreter's skin, his racial or ethnic affiliation.

The figure of the interpreter Yussuf takes centre stage in the almost symmetrical composition of the photograph, emphasizing his pivotal role as "middleman". The straightforward view, along with the frontal positioning of the body implies a situation of face-to-face interpreting. But, since the eyes of the interpreter are shadowed as a result of the strong sunlight, leaving black eye hollows in a dark, only patchily illuminated face, eye contact as part of the face-to-face communication is obviated. This impression is reinforced by the interpreter's very serious, motionless, even unapproachable facial expression together with the stiff or frozen, slightly strained posture of his upper body. Only his legs have moved a little out of the straight body line, opening up in the direction of the photographer/foreign visitors. Interaction seems to be petrified in this figuration of the interpreter. The visualization stands in stark contrast to the interpreter-mediated interaction function which Yussuf fulfils in his social role as a native escort interpreter. Yussuf is represented as a gatekeeper, barring the way to the village, which is recognizable in the far background by the little huts beaded along the line of horizon. With his positioning in the foreground of the image – near enough, but not too close to the placement of the photographer/viewer – he shows that he possesses the sovereign power and has the standing to decide what to show while escorting the visitors and what to communicate to them during the process of interpreting.

The position of the adolescent interpreter Yussuf is that of an influential cultural broker. He acts as a spokesman with a strong mandate. This is reflected not only in his body posture, but also in his dress. The full-body portrait represents Yussuf in a semi-military look. His jacket resembles a military uniform, with its continuous button facing and the stand-up collar. One further indicator for a general's uniform and an emblem of Western modernity, defined by the acceleration of time, is the pocket watch whose chain is affixed to the button border. On his head, Yussuf proudly wears a fez, a symbol of Ottoman modernity throughout the Arab world of the Near East and Northern Africa, insignia of high military or civic rank, and marker of Oriental cultural identity. The military significance of the fez is evidenced by its use as military headgear. From the late nineteenth century onwards, the fez became a typical headdress of "native" soldiers, locally recruited by colonial troops. If we compare the Abyssinian interpreter's upper-body garments with those of Sultan Yussuf Izeddin-Effendi, the crown prince of the Ottoman Empire from 1909 until his death in 1916, in another photograph from the George Grantham

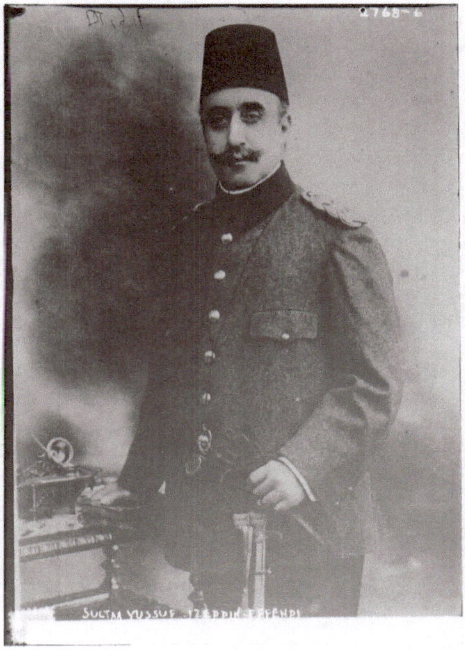

FIGURE 5.2 Sultan Yussuf Izeddin-Effendi

Bain collection (Figure 5.2),[5] it is obvious that they emulate the Ottoman sultan style of a general's uniform. However, they do not seem truly official and prestigious but rather casual, because of the mixing of elements of a uniform jacket with the style of labourer or prison clothing. Together with the native dress style of the baggy trousers, the linen kaftan shirt peeping out from beneath the jacket, and the bare feet, the combined outfit amounts to a native-adapted version of Ottoman modern dress.

The enhancement of the native adolescent interpreter's status is performed not only through up-dressing, but also through professionalization. This becomes manifest when viewing the full-body portrait of Yussuf in the context of the photo series it belongs to. As mentioned, the photograph of Yussuf was taken by a photographer working for the Bain News Service.[6] During the visit to the Abyssinian village, other photographs of important interlocutors, or informants, were produced as well.[7] The serial nature of the photographs can best be reconstructed by comparing in Figure 5.3 the photographs of Booloo, the chief camel man, and Yussuf, the interpreter.

The location where the two portrait photos are shot is one and the same, although the angle, perspective and framing of the pictures differ slightly.[8] Given this serial picture context, it becomes apparent that the adolescent interpreter Yussuf figures among high social and professional ranks. He is represented as a central figure and important actor of his native community, a professional among

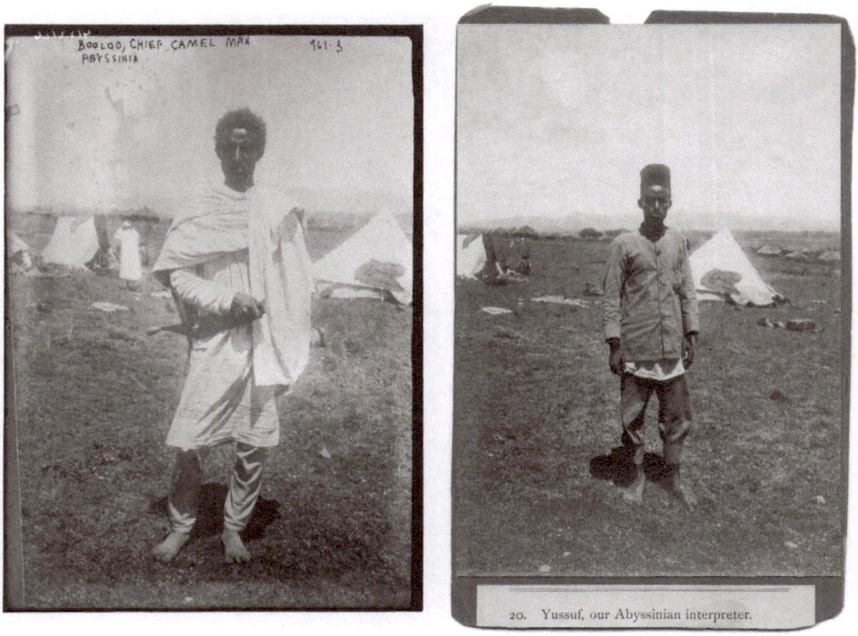

FIGURE 5.3 Booloo, chief camel man, Abyssinia (left). Yussuf, the Abyssinian interpreter (right)

other professionals, worthy of being portrayed individually in his "official" function as chosen interpreter. At the same time, the photographer does not permit him to be portrayed as an individual subject, but only as an office-holder and servant of the delegation. The statue-like posture and frontal perspective underpin this objectifying, functional approach.

In spite of the official elevation and empowerment of Yussuf to a high-ranking professional through the serial framing, his social and cultural identity is represented as broken, shifting between professionalism and unprofessionalism, maturity and naivety, modernity and aboriginality, a civilized and uncivilized way of life. The photographic configuration of the Abyssinian interpreter becomes a symbol for cultural brokerage and societal translation. Particularly in contrast with Booloo, the chief camel man, and Haji Ali, the Somali priest, who appear in front of the camera in traditional Abyssinian dress, the outfit of the interpreter Yussuf tends to be identified more with a modern secular or even military dress style as introduced by the Ottoman Empire. The appearance of the interpreter is split in bodily terms, brokering between exterior and interior, upper and lower, self-authorization and incapacitation. This is symbolized in the native barefoot dress in the leg region and the modern uniform dress of the torso. The clothing of the upper body looks slightly inauthentic, as if Yussuf has specially put on his Western-style Ottoman uniform dress with the fez for his professional activity as escort interpreter – like children who try out new social and professional adult roles by masquerading. This role-playing game of children seems to characterize the situation of the adolescent Abyssinian interpreter. He is acting out role conflicts in the form of role switching from an adolescent to an adult person, from a non-professional to a professional, from an incompletely acknowledged member of society to one with a high-ranking social status, in short: from a minor to a major (in the legal, but also the military sense).

These role changes can also be identified in the situation of colonial interpreting. Yussuf's position in the "field of translation" reflects the brokering of the colonial situation in his land and time: that is, the power relations between white Westerners and black Ethiopians.[9] He stands very close to the photographer (most likely a white American or European) in the settlement zone of the visiting foreigners, recognizable by the white tents in the middle ground. The native village lies far behind him in the liminal zone of the horizon. The anonymous photographer has taken the photograph of the Abyssinian interpreter Yussuf in his own field of translation and protected zone. The black interpreter is fully transferred into the area of the white settlement, not allowing for any bilateral or in-between situation. For the on-site shot, the photographer has chosen to position Yussuf between himself and the delegation's white tents (as in all other portrait photographs of the series), in other words clearly *on his side*, and not, as would also have been possible, between the visitors' zone and the indigenous village, or directly in front of the native community. This clearly marks the side for which the photographer expects the interpreter to stand in and interpret. The social and military empowerment of the minor interpreter into a major, indecisively oscillating between native child soldier and general, is kept under control in the area of the colonial powers' order, their translational zone of political, cultural, social and economic interests.

The interpreter of the interpreter: The power relations between the photographer and his minor subject

How can we interpret the photographic interpretation of interpreters? According to Graham Clarke,

> [e]very photograph is not only surrounded by a historical, aesthetic, and cultural frame of reference but also by an entire invisible set of relationships and meaning relating to the photographer and the point at which the image was made. … The image is as much a reflection of the "I" of the photographer as it is of the "eye" of the camera.
>
> *(Clarke, 1997, p. 30)*

Drawing on this assumption, what has inspired the photographer to visually interpret a minor interpreter? Is his aim to give centre stage to the usually invisible, servicing and humble translator? Is it to empower the interpreter by spectacularizing his in-between role? Or is it to raise general awareness of the infantilization and minoritization of interpreters, their disregard?

The photographer of the Abyssinian interpreter reveals a primary ethnographic interest in displaying a native or "natural" translator. As his subject, he has photographed a native interpreter, too young and minoritized to be trained as a professional or fully recognized as an official translator. Original or aboriginal translation, that is, genuine, untrained and unfeigned translation, is the focus of the photographer's attention. The native minoritization of the act of interpreting suggests that translation is an innate skill.[10] The neutral and transparent medium of analogue photography promotes this effect of naturalizing the interpreter. Through its natural sign language (the photograph carries the referent with it), it helps to untranslate translation, to eliminate the in-between times and spaces of translation, the formalization of the actor and of the interpreting process, and the individual personification and cultural or habitual codification of the face-to-face translational interaction. Photography presents the moment, not the process of interpreting. It displays the "here and now" of on-site, ad-hoc interpreting, although it is always inscribed in the past.

At the same time, the nativeness of the interpreter figure, his nudity and crudeness – incarnated in the display of his bare feet – seem to have a somehow disconcerting effect on the photographer, who belongs to the visitors' delegation from the civilized Western world and its colonial power. Despite the complicity and alliance with the native interpreter, the admiration for his naturalness, the nature of the "savage" child interpreter has to be domesticated. This taming of unbridled interpreting is achieved by the visual framing of the Abyssinian interpreter, his stylistic photographic codification. The white gaze of the photographer seeks to keep the black interpreter at a distance and control him. This is done using two strategies: first by positioning him at a certain distance from the photographer's own standpoint within the protected visitors' zone, the secure place of sovereign interpretation outside the native village; and second by upgrading and professionalizing the un- or semi-professional minor, that is by classifying the adolescent native interpreter as a cultural-professional group member and official representative by means of a professional type photography characteristic of social portrait photography in the 1920s.

While the photographer acts as a participant observer of his subject, serving as both his accomplice and his controller, the minor interpreter in front of the lens is presented as a participant agent, brokering between nativeness and civilizedness, blackness and whiteness, autonomy and heteronomy. It seems not entirely coincidental that the adolescent interpreter partially participates in the dress code of the dominant colonial culture of the modern West. By assimilating the power language of Western modernity, its imperial elite habitus, he looks quite "civilized" and educated. This mimicry is part of the self-empowerment of the minoritized interpreter, the desired elevation of his non- or ill-defined subject, social and professional status. However, as the post-colonial theoretician Homi Bhabha has pointed out, mimicry is a strategy of colonial power and knowledge: "colonial mimicry is the desire for a reformed, recognizable 'Other', as a subject of difference that is almost the same, but not quite" (Bhabha, 1994, p. 122). Most importantly, it articulates a metonymical relationship between self and other:

> mimicry is like camouflage, not a harmonization of repression of difference, but a form of resemblance, that differs from or defends presence by displaying it in part, metonymically. Its threat, I would add, comes from the prodigious and strategic production of conflictual, fantastic, discriminatory "identity effects" in the play of a power that is elusive because it hides no essence, no "itself".
>
> *(p. 90)*

Applying this concept of mimicry to the situation of colonial interpreting reveals that the relationship between the native child interpreter and the non-native colonial photographer is non-transparent, visibly marked by the translation of conflicting identities and loyalties. In this case, the translation of the native is broken through assimilations and mimicry. There is no original, but only resemblance, to be found in the layers of translation. By photographically interpreting the interpreter, his masquerades and camouflages, the reality effect of photography turns into an unreality effect. Assimilating the mediator status of the minor interpreter who lacks the autonomy and authority of a sovereign subject, photography indicates no clear identity as a medium of transfer. Its status shifts between documentation and imagination, veracity and illusion, truthfulness and deception. The ambivalence of the colonial discourse, its "double vision",[11] is captured in the colonial photographic framing of the minor interpreter. Colonial photographic translation and minor interpreting reflect each other: the photographer admits only a partial, inappropriate representation and recognition of the child interpreter empowered through his role as cultural broker. This way, he secures his colonial authority and sovereignty to interpret the interpretation of otherness in his own perspective.

Notes

1 For the discussion on norms, status and habitus relevant to the professional group of translators, see Sela-Sheffy, 2005.
2 See Faulstich Orellana, 2009 and Antonini, 2010 on child language brokering.
3 Escort interpreting is a "type of interpreting done when accompanying visitors, diplomats and businessmen to meetings and negotiations" (Harris, 1983, p. 5). In the context of this photograph, the interpreter has most likely accompanied a press delegation.
4 The term "thick visual translation" is used with reference to Kwame Anthony Appiah's notion of "thick translation" (Appiah, 1993/2000). It aims to highlight the situated, context-dependent visual translation of the figure of the interpreter by including the media-specific and institutional dimension of photographic translation.
5 The photograph is dated between 1909 and 1916.
6 The Bain News Service was one of America's first news picture agencies. It was established in 1898 by George Grantham Bain, a pioneering news photographer based in New York. The Bain News Service collected photographs of worldwide news and distributed them to several newspapers. The photographs became part of the George Grantham Bain Collection, which was purchased by the Library of Congress, Washington, in 1948.
7 In the digital archive of the Library of Congress, I found three more photographs by the Bain News Service that can be directly linked to the Abyssinian field trip of the

documentary photographer and his press/expedition team: a photograph of Argou, the chief of escort, one of Booloo, the chief camel man, and one of Haji Ali, a Somali priest. All three men probably worked as intermediaries for the team.

8 This can be seen from the setting, in particular the white tents in the middle ground, which show the same placement of cloths over the tent roof and on the grassed area. Even the native man who sits next to the tent on the left of the picture with a stick in his hand (probably a guard) can be seen in both photographs, albeit partly hidden by a person dressed in white in the photograph of Booloo. The photograph of the Somali priest Haji Ali has a similar setting. Despite its different background, the photograph of the chief of escort Argou can easily be assigned to the series due to the numbering in the upper right corner. All three pictures have the same handwriting in their title inscription and are numbered consecutively from 961–3 to 961–5. The photograph of the Abyssinian interpreter Yussuf is not numbered and also not part of the George Grantham Bain Collection, indicating that it somehow fell out of the series and followed other paths of distribution and archiving. One original print of the shot used in this paper was found by the ICOTI research collection on the second-hand collectibles market.

9 Historically, Abyssinia, the Empire of Ethiopia, was part of the "Scramble for Africa" headed by the European colonial powers of Britain, France and Italy. At the time when the photograph of the Abyssinian interpreter was taken (around 1925), Abyssinia enjoyed independence, but the colonial context and threat was present at all times. For these political and historical reasons, the photograph of the Abyssinian interpreter must be discussed within the broader framework of colonial photography.

10 A classic translation studies approach to theorizing "natural translation" is found in Harris and Sherwood, 1978 and Blasco Mayor and Jiménez Ivars, 2011.

11 In "Of Mimicry and Man", such double vision is directly associated with the mimicry of the colonial subject: "The *menace* of mimicry is its *double* vision which in disclosing the ambivalence of colonial discourse also disrupts its authority. And it is a double vision that is a result of what I've described as the partial representation/recognition of the colonial object" (Bhabha, 1994, p. 88).

References

Antonini, R., 2010. "The study of child language brokering: past, current and emerging research." In: "Child Language Brokering: Trends and Patterns in Current Research." Special issue of *mediAzioni*, 10, pp. 1–23. http://mediazioni.sitlec.unibo.it/ [accessed 7 April 2014].

Appiah, K. A., 1993/2000. "Thick translation." In: L. Venuti, ed. 2000. *The Translation Studies Reader*. London: Routledge. pp. 417–29.

Bhabha, H. K., 1994. *The Location of Culture*. London: Routledge.

Blasco Mayor, M. J. and Jiménez Ivars, A. eds., 2011. *Interpreting Naturally: A Tribute to Brian Harris*. Berne: Peter Lang.

Chesher, T., Slatyer, H., Doubine, V., Jaric, L. and Lazzari, R., 2003. "Community based interpreting (CBI): the interpreters' perspective." In: L. Brunette, G. L. Bastin, I. Hemlin and H. Clarke, eds. *The Critical Link 3: Interpreters in the Community*. Amsterdam: John Benjamins. pp. 273–92.

Clarke, G., 1997. *The Photograph*. Oxford: Oxford University Press.

Faulstich Orellana, M., 2009. *Translating Childhoods: Immigrant Youth, Language, and Culture*. New Brunswick, NJ: Rutgers University Press.

Gentile, A., Ozolins, U. and Vasilakakos, M., 1996. *Liaison Interpreting: A Handbook*. Victoria: Melbourne University Press.

Harris, B., 1983. "There is more to interpreting than conference interpreting." *InformATIO: Association of Translators and Interpreters of Ontario*, 11(3), pp. 4–5.

Harris, B. and Sherwood, B., 1978. "Translating as an innate skill." In: D. Gerver and W. Sinaiko, eds. *Language, Interpretation and Communication*. New York: Plenum Press. pp. 155–70.

Mason, I. ed., 2001. *Triadic Exchanges: Studies in Dialogue Interpreting*. Manchester: St Jerome.
Mikkelson, H., 1996. "Community interpreting: an emerging profession." *Interpreting: International Journal of Research and Practice in Interpreting*, 1(1), pp. 125–29.
Roberts, R. P., 1997. "Community interpreting today and tomorrow." In: S. E. Carr, R. P. Roberts, A. Dufour and D. Steyn, eds. *The Critical Link: Interpreters in the Community*. Amsterdam: John Benjamins. pp. 7–26.
Sela-Sheffy, R., 2005. "How to be a (recognized) translator: rethinking habitus, norms, and the field of translation." *Target*, 17(1), pp. 1–26.

6

FRAMING THE INTERPRETER'S WIFE

Anxo Fernández-Ocampo and Michaela Wolf

There is a small group of postcards, representative of the orientalizing aesthetics of Indochinese women, devoted to the wives of native interpreters from colonial French Indochina, all published between 1904 and 1910 by the first professional photographer in Tonkin: Pierre-Marie Alexis Dieulefils (1862–1937). Four of these pictures, along with a fifth depicting a male interpreter, are reproduced in this paper from original prints held in the ICOTI collection. The portraits do not partake in the natural history projects which were characteristic of late-nineteenth-century visual anthropology (for example, Batut, 1887); they therefore have a special value as an ethnographic record of the field of interpreting, which may help us to understand the social standing of the interpreter in colonial times, in a period of history when government and law rested on relationships of colonial power.

From the eighteenth century onwards, the French colonial administration set out in detail the duties, competences and working conditions of government interpreters and native public employees working in translation and interpreting, which in French Indochina formed part of a sophisticated native bureaucracy. This high degree of regulation reached its peak during the second half of the nineteenth and the first third of the twentieth century (Aldrich, 1996). The public journals from this period indicate the native interpreters' rights and obligations, pay scales and position in the hierarchical structure, allowing comparisons to be made between different practices in various civil and military organizations: colonial interpreting, diplomacy, foreign affairs and home affairs (for examples, see Duvergier, 1901). French Indochina was a hothouse for the colonial postcard industry, which included a significant number of interpreters portrayed as a specific "type". Female interpreters are almost entirely absent in the iconography of postcards more generally prior to the Second World War. However, some printed postcards and engravings of the industrial age use images of women to highlight particular characteristics of the linguistic mediator, and specifically make the woman an "accessory" in their representation of the male interpreter. Pierre Dieulefils contributed actively to enhancing the role of the native mediator by picturing individuals or groups of male interpreters, but this paper will examine three postcards that he published on the subject of the interpreter's wife. In them, although the interpreters' wives are not shown participating in the interpreting process, they physically occupy a space that is significant for the public image of the male interpreter.

The fact that these images of women were published emphasizes their position as a complement or counterpoint to the male interpreter and as an element of visual balance. By transferring the properties of the interpreter onto the woman or wife, the act of photography seems to reproduce the existing power relationships, as a strategy to fix the standing of the native colonial interpreter in society. In this way, some of the external objective structures of the field of interpreting are framed in the photograph. This article attempts to explain how the publication of postcards picturing the interpreters' wives contributed to the construction of that standing.

Habitus and translation studies

Applying Bourdieu's notion of habitus to the photographs examined here (Bourdieu, 1993), we may say that the interpreter's habitus is shaped by ascriptions made by the agents involved in the photographic situation. Those agents include the photographer, who uses technology to reproduce the habitus by reinforcing or undermining traditional views of the interpreter. In terms of the distinction postulated by Daniel Simeoni (1998), this study refers not to the habitus as the interpreter's capacity to acquire or produce knowledge, but to the *social* habitus – related to the interpreter's position in his or her field, which partly results from his or her relations to other agents – and more specifically the way in which that habitus shapes the interpreter's social image. From this point of view, external social agents contribute fundamentally to the definition of an interpreter's social position. This is also the perspective adopted by many recent studies of community interpreting (Angelelli, 2004; Bahadır, 2004; Inghilleri, 2005), which have turned to ethnographic methods to study interactions involving interpreters and other social agents associated with the act of interpreting.

Moira Inghilleri (2005) adopts Bourdieu's theory of the "zone of uncertainty" to describe the inconsistency and fluctuation of the field of interpreting, comparing this field to more prestigious socio-professional activities such as the work of judges, doctors and teachers – who all demonstrate and reproduce the objective structures of their respective fields (Inghilleri, 2006, p. 61). We assume that interpreting's zone of uncertainty may correlate with the zones of (un)certainty of the other duties and professional activities of the colonized society; this is why, although we agree with Inghilleri's approach, we suggest that the sense of indeterminacy she identifies is itself a characteristic of translation scholars and professionals, nourished and projected by translation studies research which does not always take into account the possible uncertainty of the other fields surrounding interpreting. In the last decade or so, sociological approaches to interpreting and translation have begun to counterbalance this shortcoming (see, for example, Wolf and Fukari, 2007; Tyulenev, 2012; Wolf, 2012).

In this paper, it is by connecting the captions with the photographs that we are able to link different practices and show the affinities between interpreting and other roles and activities of the colonized society. For example, in the corpus of postcards by Dieulefils conserved by Lionel Labastire (2004–13), the captions of the images on themes related to women highlight other social roles, such as professions ("seamstress"), economic level ("rich woman") or age ("female centenarian"), or refer to particular activities (bathing, resting, playing music, smoking a pipe). Starting from the hypothesis that the uncertainty apparently characteristic of community interpreting is not necessarily a constant or a universal (see Lahire, 2004), our claim is that the changing perception of interpreting as a discrete social field depends less on the intrinsic nature or essence of the profession of the interpreter and translator than on the particular modalities and historical contexts at stake, which result in different, and sometimes opposing, habitus forms.

The colonial world at the beginning of the twentieth century and the golden era of postcards in the industrial age offer a great deal of evidence for the existence of interpreting as a social field in that period. The captions develop just such evidence; they are designed to integrate the images of interpreters into larger series of photographs which provide a taxonomy of the types of jobs performed by the local population. Often, studies of translation and colonization view translation as a channel that runs parallel to colonization itself, forming part of education and the control of markets and institutions (Robinson, 1997). In that view, the colonial power favours the imposition and reproduction of these structures of translating. By presenting interpreting as an everyday activity across a number of areas of contemporary life, the postcard medium helped to classify the work of interpreters as a profession and to build a socio-professional category which was also clearly marked out in the nomenclature of the French colony as "interprète" (Gouvernement général de l'Indo-Chine française, 1900; 1902).

How did photography help to construct the habitus of the colonial interpreter?

The practices of systematically representing interpreters and instrumentalizing them in a range of media, especially in photographs and postcards, have undermined the invisibility usually so characteristic of the translator's habitus (Simeoni, 1998, p. 12) and conditioned their public image. Interpreters may become visible figures when designated as such through a caption to a photograph, but what the photographs themselves display are the material objects that enable the definition of the interpreter's social position. The figure of the interpreter is thus seen only through the transparent signs of his habitus. We can thus say that, for our study of interpreters' image, it is not a question of their visibility or invisibility but rather of the "transparency" of their habitus.

The archives of the French colonial government in Indochina contain important data which cast light on the elasticity or consistency of interpreting's zone of uncertainty, such as interpreters' affinity with other professional groups and the changing classification of interpreting within disparate socio-professional categories. Photography influences the mode of collective representation due to its "strong value of sociability" (Rogan, 2005, p. 18) and its externalization of perceived images, in this case images of the interpreter. It is able to make interpreting comparable to other activities and professions, revealing the porosity of the field of interpreting. At the same time, by the turn of the twentieth century the figure of the colonial interpreter was used as a means of socialization, as can be seen in the illustrated trade cards in Europe showing colonial interpreters captioned as such (Fernández-Ocampo, 2014) and targeting young audiences. The widespread use of postcards staging interpreters, and thus the projection of the interpreters' image or type to a large non-specialist audience, therefore means that the role of the interpreter is presented as a form of socialization.

Series of native types

The classificatory capacity of the postcard genre was an efficient tool in the colonial policy of presenting the local population in a closed system of typologies, a system incapable of generating new developments. Whereas in mainland France some photographs of interpreters in the style of a *carte de visite* displayed accumulated professional experience, photographs of colonial interpreters do not describe any individual position: the images are synchronic snapshots and show the interpreter's traits as inalterable. Together with serial publication, the tendency to represent identifiable and interchangeable traits – inalterability – is one of the main features of the postcard as a type of communicative ritual (Rogan, 2005, p. 16). Serialization and inalterability are both located on the level of repetition and institutionalization.

To a certain extent, the aim of publishing series of pictures of "natives" was to fix a unified morphology of the racial type of the indigenous employee in the colonial administration. The images represented a macro-social level with an observable habitus, showing the public the various types of relationship between interpreting and other professions and the relationships between the interpreter and his surroundings. In one series of literary descriptions of colonial types in Indochina by the colonial French writer Louis Peytral (1897), the chapter entitled "Les interprètes indigènes" (The native interpreters) depicts the different degrees of assimilation of European codes among Saigonese and Tonkinese native interpreters by comparing their social customs, hairstyles and ways of dressing. Peytral also addresses the marital relationship in both cases: while the lower-class Tonkinese interpreter combines all sorts of elements, including his wife, to prove his success vis-à-vis native society, the Saigonese interpreter, attributed a higher social standing, does not hesitate to parade his wife and her adornments in public, further revealing his concern for appearances by treating his wife as a precious object (Peytral, 1897, p. 119). This is the historical context for the metonymical

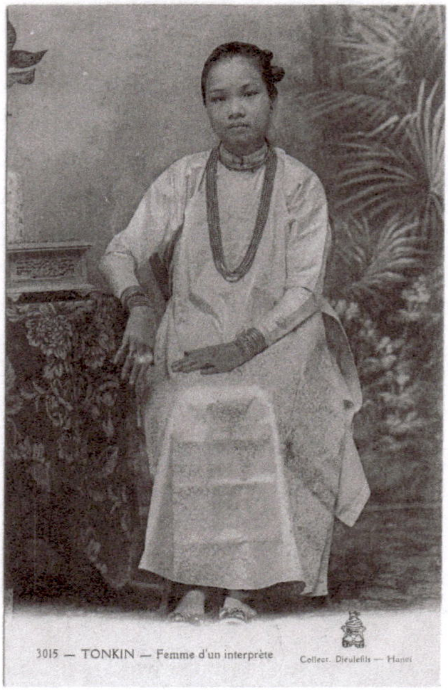

FIGURE 6.1 Tonkin. Wife of an interpreter

substitution of the male interpreter by his wife found in the photographs examined here: the native interpreter has to have the kind of wife expected of him and to ensure that she is seen publicly, in a suitably "bourgeoise" attitude. Dieulefils's postcards and artificial scenes give a glimpse of the interpreter as an owner of material property, married to a bourgeoise in all her beauty – as in Figure 6.1, captioned "3015. Tonkin. Femme d'un interprète" (3015. Tonkin. Wife of an interpreter).

Dieulefils opened a studio in 1888, soon after he settled in Hanoi. His activity as a publisher of postcards of French Indochina took place between 1902 and 1910. His postcards were sold on various markets such as colonial exhibitions, and many circulated between Europe and the colonies in the East via communication and transport networks. For this reason, they were often published in multilingual versions, as is the case here.

The bust shot (Figure 6.2) captioned "Femme d'un interprete [sic] annamite" (Wife of an Annamite interpreter) apparently comes from a photo campaign by Dieulefils. According to the caption, she is the wife of an interpreter who was working on a railway bridge construction site on the Nậm Thi River. Her hypothetical husband, or a different interpreter supposedly working on the Nậm Thi construction site, is pictured in the same series (Figure 6.3).

The young woman and man who are posing for these studio photos in Figures 6.2 and 6.3 appear to have been muses for Dieulefils, for there are complete series of studio photographs featuring the same models in a variety of poses and roles,

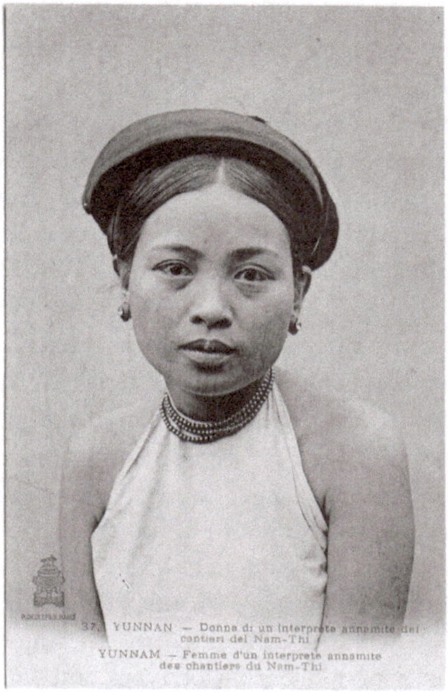

FIGURE 6.2 Yunnan. Wife of an Annamite interpreter from the Nậm Thi construction site

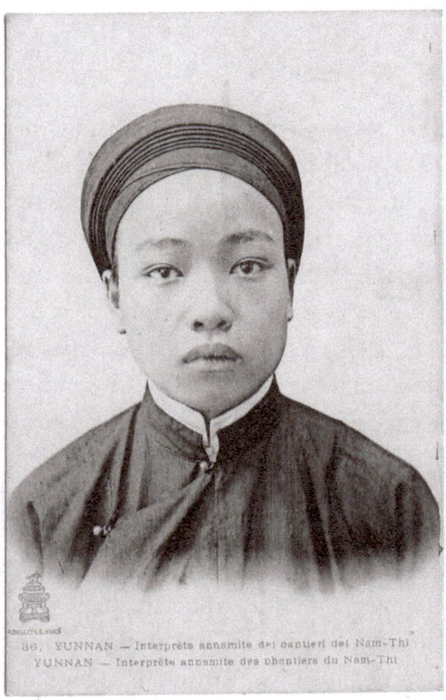

FIGURE 6.3 Yunnan. Annamite interpreter from the Nậm Thi construction site

published in different editions with different captions. Such recycling of images was common practice among postcard producers at the time. Dieulefils attributed different origins and social roles to his models each time. Thus, Figure 6.1 also exists in a version captioned "1402. Cochinchine. Femme de Saïgon" (1402. Cochinchina. Woman from Saigon) and, while Dieulefils captioned Figure 6.2 "37. Yunnan / Donna di un interprete annamite dei cantieri del Nam-Thi / Yunnam / femme d'un interprete [sic] annamite des chantiers du Nam-Thi" (37. Yunnan / Wife of an Annamite interpreter from the Nậm Thi construction site), the same image also appeared as "229. Tonkin / Haïphong. Femme" (229. Tonkin / Haiphong. Woman). In other cases, the same muse is photographed in a different pose, such as "407. Tonkin. Femme faisant la lecture" (407. Tonkin. Woman reading) (Labastire, 2004–13). The same photograph of the male model (Figure 6.3) was also published as a postcard with the caption "188 Tonkin. Phu Lang Thuong. Jeune homme" (188 Tonkin. Phu Lang Thuong. Young man). This manipulative practice underlines the centrality of captions in the production of meaning, and for our purposes specifically in managing the field of interpreting. At the same time, it indicates that the designation "interpreter" is no accident, but a deliberate choice by the postcard editor: Dieulefils is committed to ensuring that the figure of the interpreter is included in a periodic table of types and functions of the French colony.

However, the photographic dialectic between the supposed wife and the interpreter can yield a quite different interpretation once other elements of the picture's composition are analysed, as will be shown in the next section.

Habitus and material possessions

The carefully composed photograph in the two versions of Figure 6.4, captioned "Femme d'un interprète dans son intérieur" (Interpreter's wife in her living room) and showing a "wife" in an exoticizing pose, exemplifies the continuum between – and reciprocal legitimation of – studio photography and ethnographic photography. The photographer has included what he considers to be pertinent objects to compose an observable framework with the technique of visual inventory (Collier and Collier, 1986, pp. 44–63). Dieulefils recreates the bourgeois aesthetic of the cloistered woman, reinforced by the symbolic attributions that the wife of an interpreter is supposed to have. The two versions of the scene come from the same session and were published by Dieulefils under the same series number. In both cases the wife is shown holding her husband's notebook in her hand. To facilitate interior photography, where the scarcity of light allows only limited movement, the woman lies on her side, her knees slightly bent, in a pose which the European public would easily associate with that of an opium smoker. Against a backdrop of perpendicular lines, the oblique lines indicate a continuum between the woman and the musical instrument. Her bent leg and hips reflect the shape of the body of the instrument hanging on the wall and the direction of its neck.

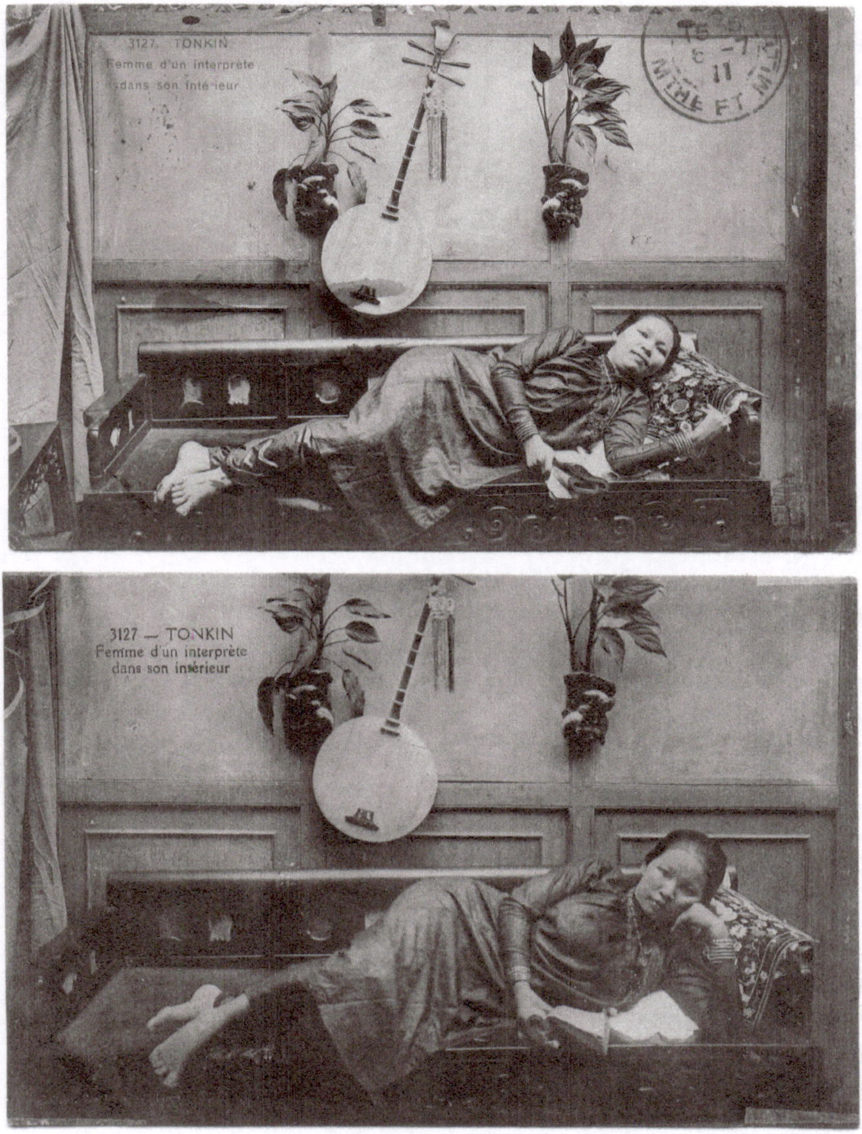

FIGURE 6.4 Tonkin. Wife of an interpreter at home

Although few in number, Dieulefils's photographs of native interpreters' wives in the French colonial administration in Indochina allow us to study the interpreter's socio-professional habitus by showing him as part of a family unit. Portraying his domestic material possessions, the photographer integrates him into a domestic continuum and ties him to particular historical and spatial coordinates. In his written account, Peytral too highlights the importance of family to the status of the interpreter, claiming that the Tonkinese interpreter prefers to receive his native clients in his home, in his family space. In that space, the interpreter's wife discreetly encourages her husband to interpret and "with her honeyed voice and outstretched hand, emphasizes like an echo each new sentence" (Peytral, 1897, p. 122; our translation[1]). On the same lines, Dieulefils's photography seems to signify that the figure of the wife is a metonymic allusion to her husband, the interpreter, in such a way that the Western observer would easily identify the wife with him.

Since in colonial discourse the woman was commonly viewed as the bearer of the man's social condition, photographing the interpreter's wife in her living room was a powerful way to project the social condition of the interpreter, physically embodying him as a woman. Although Dieulefils also pictured several male interpreters, his photographs of women reinforced the idea that interpreting is a representational profession. This supports Bourdieu's finding that professions involving presentation and representation (such as sales or marketing) and concerned with the distribution of symbolic goods were characteristic of the growth of new bourgeoisies (Bourdieu, 1979, p. 415). Through actions external to the photographic image, such as adding printed captions, the postcard producer calls for a *nouveau bourgeois* model to locate the native interpreter in the social spectrum of French Indochina. As a privileged means of communication, the colonial postcard industry thus played a key role in showing how native interpreters should perform their duties within their own society. As an external social agent, the photographic editor pictured these interpreters with European attributes and thus provided them with a biography in which to ground their professional habitus. At the core of these biographies were the wives, who were presented in such a way as to fit into the expected habitus.

Towards the creation of a public image

In the postcards discussed in this study, there are no visible signs of the process of interpreting. The public would be unable to identify the interpreter in postcards using portraits alone, so that captions are vital in order to describe and regulate a socio-professional category and associate it with a human figure. The use of the caption "interpreter" can be found in the published colonial photographs and also in amateur photography (see, for instance, the portrait of Yussuf discussed by Mersmann in this volume). Even more revealing than his decision to picture the wives rather than the male interpreters is the fact that Dieulefils did not even need to shoot a real wife. He was operating from existing, internalized images of each socio-professional type of the colonial population. The specific conditions of the photographic studio allowed him to shape and reshape this image at will by giving a generic face to every occupational group, hence inserting the visual category of the colonial interpreter into the "type series" of Indochina.

Our analysis has suggested that the practice of photography in colonial Indochina helped to produce the idea of a social field of interpreting, and that the portraits of interpreters and the written captions designating their status are external actions which contributed to the creation of recognizable professional schemata for the general public. These external images may also have enabled the collective of native interpreters to legitimize their profession. In addition to the captions which institutionalize the role of the interpreter, the display of clothing, objects and settings enhances the social normality of the interpreter in visual culture, making him a recognizable figure in the urban and semi-urban colonial landscape and regulating his public image through his marital status. Furthermore, the accumulation of external signs of wealth and social success, as part of the *illusio* in Bourdieu's terms, suggests that the status of the interpreter correlates with the act of assuming professional responsibility as reflected in social success.

By captioning studio photographs with the category "the interpreter's wife", Dieulefils imbued the figure of the interpreter with an aesthetic and rhetorical value for French colonial imagery. As a social sub-category in the typological representations of French Indochina at the start of the twentieth century, the interpreter's wife enriches our view of the modalities of the social field of interpreting and offers the chance to study the interpreter's public image in the colonial context from the perspective of translation studies.

Note

1 "la voix mielleuse et la main tendue, souligne, comme un écho, chaque phrase nouvelle".

References

Aldrich, R., 1996. *Greater France: A History of French Overseas Expansion*. London: Palgrave Macmillan.

Angelelli, C. V., 2004. *Medical Interpreting and Cross-cultural Communication*. Cambridge: Cambridge University Press.

Bahadır, Ş., 2004. "Moving in-between: the interpreter as ethnographer and the interpreting-researcher as anthropologist." *Meta*, 49(4), pp. 805–21.

Batut, A., 1887. *La photographie appliquée à la production du type, d'une famille, d'une tribu ou d'une race* [*Photography Applied to the Production of a Type, of a Family, of a Tribe or of a Race*]. Paris: Gauthier-Villars.

Bourdieu, P., 1979. *La distinction* [*Distinction*]. Paris: Les Éditions de Minuit.

——, 1993. *The Field of Cultural Production*. Translated by R. Johnson. Cambridge: Polity Press.

Collier, J. and Collier, M., 1986. *Visual Anthropology: Photography as a Research Method*. Albuquerque: University of New Mexico Press.

Duvergier, J. B. ed., 1901. *Collection complète des lois, décrets, ordonnances, règlements et avis du Conseil d'État* [*Complete Collection of Laws, Decrees, Orders, Rules and Notices of the State Office*]. Vol. 101. Paris: Librairie de la Société du recueil général des lois et des arrêts.

Fernández-Ocampo, A., 2014. "Le déni du métissage: l'interprète et l'espace dans la gravure coloniale" ["The denial of métissage: space and the interpreter in colonial engraving"]. In: J. Yuste Frías, ed. "Métissages et traduction." Special issue of *Alterstice*, 4(1). www.alterstice.org [accessed 7 April 2014].

Gouvernement général de l'Indo-Chine française, 1900. *Cochinchine: budget local pour l'exercice 1901* [*Cochinchina. Local Budget for the Financial Year 1901*]. Saigon: Imprimerie coloniale.

——, 1902. *Compte administratif du budget local du Tonkin. Exercice 1901* [*Administrative Account of the Local Budget of Tonkin: Financial Year 1901*]. Hanoi: Imprimerie typo-lithographique F.-H. Schneider.

Inghilleri, M., 2005. "Mediating zones of uncertainty: interpreter agency, the interpreting habitus and political asylum adjudication." *The Translator*, 11(1), pp. 69–85.

——, 2006. "Macro social theory, linguistic ethnography and interpreting research." *Linguistica Antverpiensia*, 5, pp. 57–68.

Labastire, L., 2004–13. *Pierre Dieulefils: Photographe de l'Indochine* [*Pierre Dieulefils: Photographer of Indochina*]. www.pierre-dieulefils.com [accessed 7 April 2014].

Lahire, B., 2004. *La culture des individus: Dissonances culturelles et distinction de soi* [*The Individual's Culture. Cultural Dissonance and Distinction of the Self*]. Paris: La Découverte.

Peytral, L., 1897. *Silhouettes tonkinoises* [*Silhouettes of Tonkin*]. Paris: Berger-Levrault.

Robinson, D., 1997. *Translation and Empire*. Manchester: St Jerome.

Rogan, B., 2005. "An entangled object: the picture postcard as souvenir and collectible, exchange and ritual communication." *Cultural Analysis*, 4, pp. 1–27.

Simeoni, D., 1998. "The pivotal status of the translator's habitus." *Target*, 10(1), pp. 1–39.

Tyulenev, S., 2012. *Applying Luhmann to Translation Studies: Translation in Society*. London: Routledge.

Wolf, M., 2012. "The sociology of translation and its 'activist turn'." *TIS Translation and Interpreting Studies*, 7(2), pp. 129–43.

Wolf, M. and Fukari, A. eds., 2007. *Constructing a Sociology of Translation*. Amsterdam: Benjamins.

PART III

First World War

Messengers of victory

7

STAGING THE ENTENTE IN THE FIRST WORLD WAR

Michael Kelly

The Entente Cordiale was only ten years old when the First World War broke out. Britain and France had come close to war over their conflicting ambitions in Africa, but after the Fachoda Crisis of 1898 they both worked to build a close relationship, which was put to the test at the outbreak of war in the summer of 1914. The stakes were very high and close cooperation between the two countries was a military and diplomatic necessity. It was vital that such cooperation should be clearly visible to the populations of the two countries and to other countries. As a result, the two governments were careful to present the picture of a close relationship, particularly between the two armies. They made every effort to "stage" the Entente Cordiale, making it visible in the public statements of political and military representatives. The control of information, including images, was an important part of the war effort, and every opportunity was taken to emphasize the strength of cooperation through visual representations. This is clear from the photographs that were published.

Military cooperation depended on the forces being able to communicate effectively with each other. Most of the work of communication fell to interpreters, whose role was to ensure that the two sides could understand each other and to smooth relations between them. Some specialist interpreters provided interpreting at high-level meetings, but for the most part it was general-purpose interpreters who carried out communication and liaison tasks on the ground. Most were French soldiers attached to British units,[1] though some were civilians.

How interpreters are depicted is particularly complex. On the one hand, as Kumiko Torikai points out, the role of interpreters is most commonly to be quasi-invisible, serving as intermediaries between actors in events rather than as actors in their own right (Torikai, 2009). She compares them to the *kurogo* in Japanese Kabuki theatre, a figure dressed in black that unobtrusively handles the actors' props, while conventionally being ignored by the audience.

The interpreter has a solitary role, generating a "sense of loneliness" that has been described by Mikhail Gorbachev's interpreter, Igor Korchilov (1997, p. 22). This was compounded in the First World War by the fact that the military interpreter was most frequently amid foreign troops. While the interpreter provided a link, the person to whom he was responsible was usually not present.[2] The interpreter was "attached" to the unit he was serving, but not a full member of it. In that sense interpreters were marginal. To minimize the sense of alienation, Stephen Bochner argues that interpreters need "social support; including a clearly defined professional role with a reasonable degree of status, and a reference group" (Bochner, 1981, p. 32).[3] Certainly, the French interpreters had a defined role and a degree of status. They attended training programmes and they were mostly working in their own country. Once they were deployed, however, they were generally without a reference group.

Fighting heroically side by side

The images circulating in the public media took many forms, including cinema, painting, illustrations and cartoons, but the most widely circulated were photographs, four of which are studied here. They were cheap to produce and easy to reproduce as postcards. As a result, photographs of interpreters circulated publicly. The first example (Figure 7.1) is of a front-page photograph in the popular

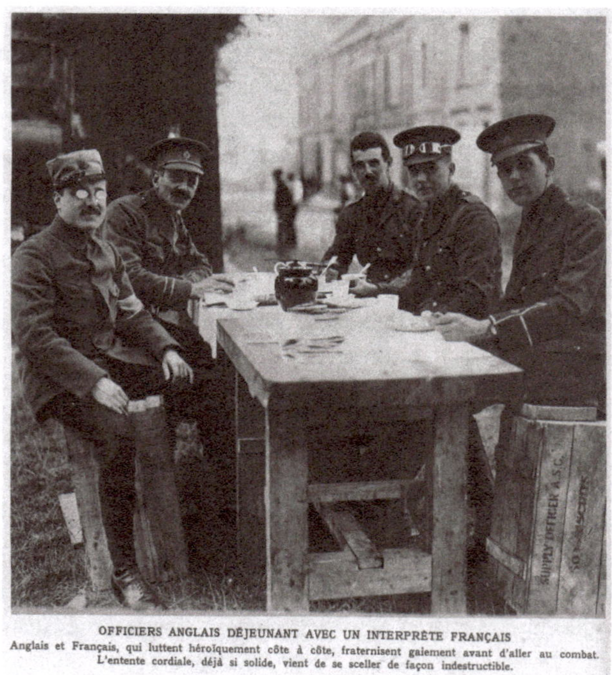

FIGURE 7.1 British officers at lunch with French interpreter

illustrated weekly magazine *Le Miroir* of 6 September 1914, where a French interpreter is depicted sitting at a table with a group of British officers.

Le Miroir was a highly patriotic magazine, which specialized in depicting the war (Albert and Terrou, 1988). As the masthead announces, the paper would "pay any price for photographic documents relating to the war, presenting items of particular interest".[4] The photographs were chosen to confirm the paper's editorial focus, and this image was certainly selected to portray the cooperation between the British and French armies.

In a magazine, images are rarely left to speak for themselves, but are channelled to ensure that they convey the meaning intended by the editors. The main mechanism for controlling meaning in photographs is the interplay of meaning between the image and its accompanying text. In press photographs, as Roland Barthes argues, the caption operates to introduce a clear connotation of what the image means, and "anchors" the many potential meanings that a bare image may support (Barthes, 1977, pp. 25, 40–41). In this case, the caption makes clear what this war photograph is intended to represent:

> British officers having lunch with a French interpreter. The British and French, who are fighting heroically side by side, are fraternizing gaily before going into battle. The Entente Cordiale, already so solid, has now been sealed indestructibly.[5]

This anchors the cover photograph as an exemplification of the solidarity of the two armies. The reader is likely to regard this description as unproblematic, and will see in the photograph a clear depiction of the relationships of fraternity the magazine is keen to confirm, with a French interpreter sitting round the table with British officers. In this way, the photograph offers what Barthes calls a *studium*, an image of general human interest which exemplifies a topic (Barthes, 1981, p. 26).

However, by their nature photographs convey more information than can easily be controlled, and usually lend themselves to a variety of interpretations. Barthes suggests that in some cases an element within the photograph visibly disrupts the unity of the image. This is what he calls the *punctum*: a detail or an aspect that compels the viewer to think about the context of the image rather than simply receiving an overall impression. The *punctum* in the photograph from *Le Miroir* is the interpreter's spectacles, which conceal his eyes and give him an owlish look that distinguishes him from the four officers. Wearing spectacles suggests that he is unlikely to engage in the same physically demanding action as the officers. He will probably not be "going into battle" as the caption suggests. This alerts the viewer to other differences. He wears a different uniform, marking him out as French. Later in the war, French interpreters sought closer integration, adopting the British khaki and retaining only the French *képi* (see Heimburger, 2012). He is probably not an officer, since he has no visible insignia other than an armband which identifies him as having an auxiliary function. He appears to have no place setting and it is not certain that he has actually eaten lunch with the British officers. He is leaning backwards while the others lean forward towards each other, suggesting a slight awkwardness, befitting an outsider and mediator.

The *punctum* subtly subverts the *studium* of carefully displayed official cooperation, and provides an uncomfortable reminder of the social and operational realities behind the staged event. The image somewhat belies the thrust of the anchoring caption and suggests a more complex, perhaps more disquieting reality behind the representation.

Because of their mediating position, interpreters sometimes figure as a *punctum* in the public images of encounters between the British and French military, referring the viewer to the communication difficulties and the degrees of foreignness inherent in those encounters. In this image, the suggestion of foreignness may also lead the reader to reflect on the setting. The scene is almost certainly set in France or northern Belgium, where fighting was taking place during the first battle of the Marne at the time the magazine was published. The presence of a driver, indicated by his goggles, indicates that it might be a temporary halt, perhaps before going into the conflict zone. However, the interpreter appears as the guest rather than the host at the table. The table itself is a rough but sturdy construction and the diners are sitting on empty biscuit boxes, supplied by the British Army Service Corps. It is a military installation in which the British officers are clearly at home. It may be, to paraphrase Rupert Brooke, that this is a corner of the foreign field that is at least temporarily England.[6] Although the interpreter is crucial to staging the Entente, he remains a curiously marginal figure.

Our friends, the British

The opposite of this marginality is presented in another postcard from the period (Figure 7.2), which depicts "The first British officer assigned to work with a garrison commander, guided by a French military interpreter".[7]

Here, it is the British officer who appears clearly as the outsider, while the French interpreter is among his own people, all wearing the French *képi*. The Briton lurks rather uncomfortably behind the interpreter. In this case, it is not the interpreter who provides the *punctum* but the French officer looking at the camera, in a hostile manner, as if the photographer is intruding in the event. Perhaps his suspicion is provoked by the presence of a civilian photographer.

The image is reproduced on postcards from the Provost company in Toulouse, a long-established family firm of professional photographers that published a large number of wartime photographs. It is conceivable that the presence of a commercial photographer at this event may have been unwelcome to some of the military participants.

This photograph was published as part of a range of postcards, a popular form of communication in the period. They were given the general title of "La Guerre 1914" ("The 1914 War") and appeared in several thematic series. Earlier series had included photographs of Moroccan (second series) and Indian soldiers (third series) (Photographie Provost, 2014). This image appeared in the fifth series, which was

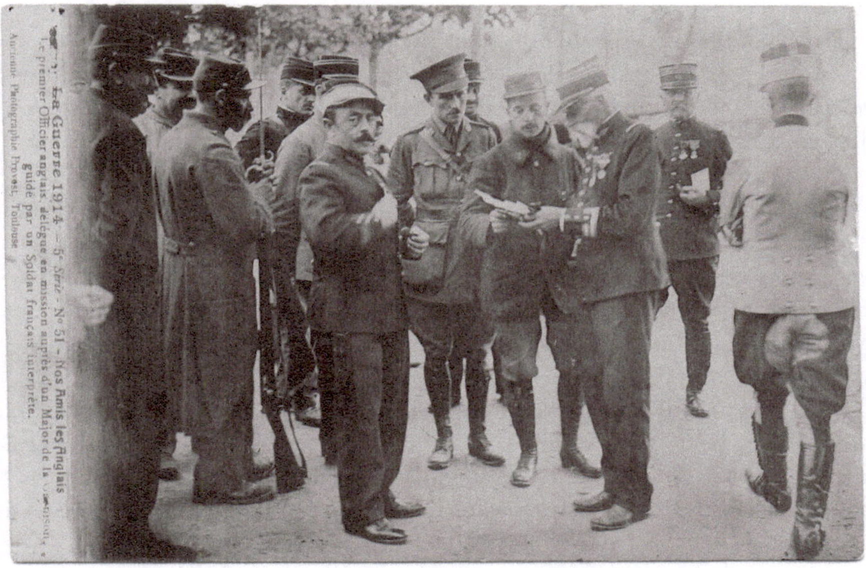

FIGURE 7.2 The first British officer assigned to work with a garrison commander, guided by a French military interpreter

given the title "Nos Amis les Anglais" ("Our friends, the British"), and it may be that the British soldiers were regarded as sharing some of the exotic flavour of the colonial troops.

The apparent hostility of one officer raises a broader question of why the photograph might be an intrusion. It may spur the reader to wonder whether the description corresponds to the events depicted. Indeed, the caption referring to "Our friends, the British" appears somewhat ironic, since there is only one British officer and he is not the focus of attention. There are no visible signs of welcome and the officer does not appear to be acknowledged by the venerable "Major de la Garnison" or by the other people present.

The caption describes the interpreter as "guiding" the British officer, but it appears here that he is instead replacing the British officer. The interpreter's uniform shows him to be part of the same army as the French participants, with a French cap and the interpreter's insignia on his collar. His stance shows that he is comfortable and fulfilling his duty, presenting paperwork to the senior officer. The Briton is looking over the interpreter's shoulder at the documents, and waiting to be acknowledged. The interpreter is not doing any interpreting, since his officer is not speaking or being addressed. He is, rather, acting as a liaison officer, carrying out formalities after which the British officer might expect to be recognized.

The signs of ceremonial in the scene confirm the sense of intrusion. The officers are wearing their medals and are engaged in interactions with each other that suggest the aftermath of a formal event. The ceremony is clearly not directed at receiving the British officer, and there is a sense that his arrival is also felt as an intrusion. The intermediary role of the interpreter, it seems, serves to secure the acceptance of the officer within the French garrison.

The disruption created by the *punctum* finally points to a difficulty within the *studium*, that there is no thematic unity within the picture. The composition is a triptych, with the central group of the major, interpreter and British officer guarded from the photographer by the French officer. The right-hand panel shows two French officers interacting with each other as if concerned with different business. The left-hand panel shows French troops at ease, one holding a rifle with its butt resting on the ground. They appear to be waiting for instructions. The caption only refers to what is happening in the central panel, where the interpreter is not only central but is also the initiator of events, presenting the orders to which the major and the British officer must then respond.

Taken together, the magazine cover and the postcard show the contradictory nature of the interpreter – a figure both marginal and central. Both aspects are vital parts of the intermediary role played by military interpreters in the wartime context, enabling them to be initiators when required and to fade into the background at other times.

The interpreter as boxer

Although interpreters largely worked alone while on duty, they also had a reference group and a rank that provided them with status. Their broader social standing is reflected in the representation of the interpreter Max Abbatt in Cope's boxer series of cigarette cards (Figure 7.3).

During the First World War, a large proportion of the male population of the combatant countries smoked, as did a small proportion of women, and cigarettes

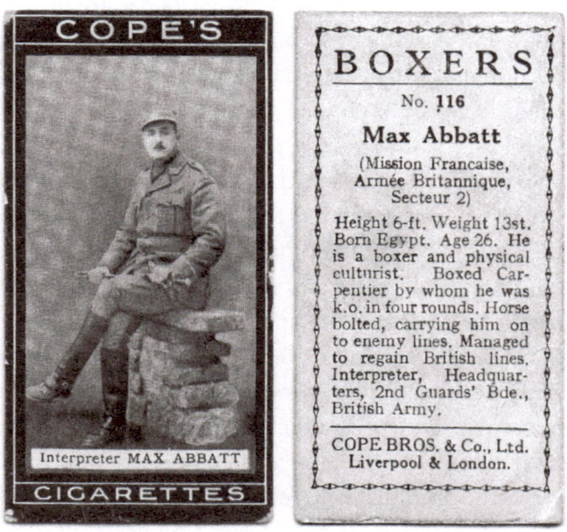

FIGURE 7.3 Interpreter Max Abbatt

had great symbolic resonance as signs of humanity and respect. To be included in a series of cigarette cards was a mark of celebrity, since the picture cards in cigarette packets were frequently collected and exchanged by both smokers and non-smokers. They are still widely collected in Britain and the United States (see Towersey, 2012; LCCC, 2014).

The card depicting Max Abbatt was part of a series of 125 cards showing boxers, issued by the British tobacco company Cope Bros. & Company in 1915 (PSA Card Facts, 2014). All of the cards represent well-known boxers, most of whom were serving in the armed forces. Abbatt is the only interpreter. About half of them are depicted in boxing shorts, while the remainder, like Abbatt, are in uniform.

The photograph does not evoke boxing. It shows a military personage in full uniform. He wears riding boots and carries a riding crop, which was a signal of status since horses were generally reserved for officers. Interpreters had to accompany officers, and therefore often rode horses themselves, benefiting from the status that this brought. Abbatt is depicted sitting on a stack of blocks that suggests the remains of a wall destroyed by bombardment and evokes the dangers of life in the army.

As a *studium*, therefore, the picture shows a military interpreter, perhaps taking a break from activities. It indicates that Abbatt's military duties are the most important aspects of his life. He is sitting alone for a formal portrait, and is not depicted in his active role as intermediary, whether at the margins or in the centre of activity. It is therefore his rank and status that are conveyed. Since it is a staged portrait, there is little scope for the kind of disruptive detail that would provide a *punctum*, though the lack of reference to boxing might suggest some discrepancy by implying a whole dimension of his life that is not depicted.

The picture of Abbatt is accompanied by different kinds of text, which serve as anchors in different ways. On the front of the card, the portrait is accompanied by his name and rank, which focus attention on the man and his military role. It is framed top and bottom by the brand and product name: Cope's Cigarettes. These suggest some endorsement by the subject, a matter to which a later reader might be more sensitive, though it was occasionally an issue even at the time, since an American baseball player famously refused to allow his picture to be used in this way.[8]

On the reverse of the card, a short text gives a summary of his personal characteristics and achievements, providing an anchor, in Barthes's sense, by clarifying which aspects of the picture the reader should attend to. His role is confirmed by his attachment to "Mission française, Armée britannique, Secteur 2" (the interpreting and liaison service), and by his posting to the Guards. The boxing connection is made explicit (Boxed Carpentier by whom he was k.o. in 4 rounds), though there is no visual sign of it in the portrait. And the broader military connection is amplified (Horse bolted, carrying him on to enemy lines. Managed to regain British lines), justifying the riding crop depicted.

For all its tweet-like brevity, the text goes beyond the function of a caption, introducing new and intriguing information that is not available in the image and projecting the reader into new areas of narrative. As a result, the functions are reversed and the image begins to serve as an anchor for the stories, each of which is presented in "taster" form. A boxing bout with Georges Carpentier (who was number 13 in Cope's boxing cards) is clearly the pinnacle of a distinguished boxing career. Surviving a horse bolting is just the hint of a dramatic military episode between British and enemy lines. In both cases the reader is left to imagine the story. Boxing championships and life-and-death drama on the front line rather eclipse Abbatt's posting to the Guards HQ. Perhaps this is another sense in which the role of the interpreter is viewed as marginal.

Armistice railway carriage

A different picture of the interpreter emerges from the final image (Figure 7.4), presented in an amateur photograph taken on the signing of the Armistice in November 1918 and depicting a specialist interpreter involved in the negotiations. The photograph shown here is a postcard reproduction that anchors the image with a detailed caption and the names of the officers portrayed.

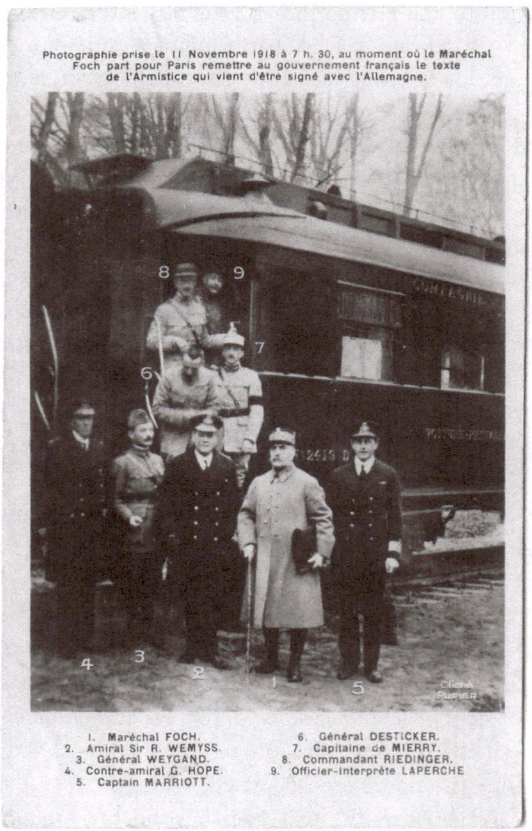

FIGURE 7.4 Armistice railway carriage

The image is one of such high symbolism that it was published in the press throughout the world and widely reproduced in postcard form. It shows the French Army Generals, the British Naval Admirals, and their staff in front of the railway carriage in which the Armistice was signed. It was taken in the moments after the German delegation had left and before Foch and Wemyss left for Paris to present the signed agreement to their governments. It is the only photograph of the successful Allied negotiators, though several artists later painted imagined versions of the signing. As a *studium*, it records the moment of victory at the end of the war. The bare details are given in the upper caption, while the lower caption lists the names of the participants. The image credit, "cliché Pupier", refers to the junior officer who happened to have a camera with him and took the photograph. The last name on the list is Officier-Interprète Laperche, who was Foch's interpreter.

The postcard reproduction helpfully provides a key to the people shown, generally giving their rank and surname. Little is known of Paul Laperche, whom some sources cite as Lieutenant (Mémorial de l'Armistice, 2014). His relative invisibility is confirmed in the photograph, where he appears as the face of a shadowy figure, peering out of the carriage door.

Laperche's highlighted face and apparent smile act as a *punctum* in the image, revealing that the war leaders, Generals Foch and Weygand and Admirals Wemyss and Hope, are dependent on the work of others who do not necessarily share in the recognition. The naval intelligence officer who provided interpreting for Wemyss, Commander Walter Bagot, does not appear in the picture at all, though his informal account of the negotiations provides an important source of information on the detail of proceedings (Voices of the Armistice, 2008).

A first reflection might be to suppose that the French and British leaders would be unable to speak to each other in the absence of at least one of the interpreters. However, Commander Bagot's account makes it clear that the interpreters were required to translate between French and German but not between French and English. The entire negotiations were conducted in French, which was adequately understood by the British side and by some of the German delegates, who also had their own interpreting service. Laperche's role was to convey in German the text of the Armistice and Foch's comments. A second reflection might be that the relative disarray in this photograph, the only one taken, betokens a lack of recognition of the magnitude of the event. This may be nearer the truth, since Bagot notes that the French general staff showed no enthusiasm for the outcome and no inclination to celebrate the moment. In this respect the excitement felt by the interpreters, now freed from their duties, was closer to the popular mood on that day.

Conclusions

Interpreters were key actors in facilitating the alliance of the two countries working together in the Entente Cordiale, but they could be awkward figures to depict in their working role as intermediaries. In staging the Entente Cordiale, governments preferred to focus on smooth and harmonious collaboration, rather than on the effort that was needed to achieve it. Depicting interpreters in photographs, whether visible or not-quite-invisible, introduced a potential *punctum*. It pointed to the effort required to stage-manage the Entente and revealed some of the ragged edges in the relationships they were mediating. The photographs we have examined show that interpreters sometimes appeared as integrating figures in the *studium* of military cooperation, which was crucial to the allied war effort and usually highlighted by the anchoring text. But sometimes, in the early days of the war, they appeared as isolated and marginal, providing a *punctum* that revealed the strains of collaboration. The photographs provide more information than can be harnessed to a linear narrative and a rich source for understanding the complexities of what they depict.

Notes

1 See Franziska Heimburger in this volume.
2 Military interpreters were invariably male, though some civilian interpreters were female.
3 I am indebted for this reference to María Manuela Fernández-Sánchez.
4 Here and throughout, translations from the French are my own.
5 "Officiers anglais déjeunant avec un interprète français. Anglais et Français, qui luttent héroïquement côte à côte, fraternisent gaiement avant d'aller au combat. L'entente cordiale, déjà si solide, vient de se sceller de façon indestructible".
In this and other cases, the French caption refers to "officiers anglais", but the translation "British officers" is adopted, since there is no apparent intention to distinguish between English, Scottish, Welsh or Irish officers.
6 See the poem "The Soldier" by Rupert Brooke: "If I should die, think only this of me:///That there's some corner of a foreign field///That is for ever England" (1914/1920).
7 "Le premier Officier anglais, délégué en mission auprès d'un Major de la Garnison, guidé par un soldat français interprète".
8 Honus Wagner insisted that his picture should not be used, as a result of which the few cards that were printed are now rare and valuable (Holleman, 2012).

References

Albert, P. and Terrou, F., 1988. "La presse en France dans la grande guerre" ["Publishing in France in the Great War"]. In: *Histoire de la presse* [*History of the Press*]. Paris: Presses universitaires de France. pp. 76–80.
Barthes, R., 1977. *Image Music Text.* Translated by S. Heath. London: Fontana.
——, 1981. *Camera Lucida: Reflections on Photography*. Translated by R. Howard. New York: Farrar, Straus & Giroux.
Bochner, S., 1981. "The social psychology of cultural mediation." In: S. Bochner, ed. *The Mediating Person: Bridges Between Cultures*. Cambridge, MA: Schenkman. pp. 6–36.
Brooke, R., 1914/1920. "The soldier." In: L. Untermeyer, ed. 1920. *Modern British Poetry*. New York: Harcourt, Brace & Howe. www.bartleby.com/103/149.html [accessed 7 April 2014].
Heimburger, F., 2012. "Of go-betweens and gatekeepers: considering disciplinary biases in interpreting history through exemplary metaphors. Military interpreters in the Allied coalition during the First World War." In: B. Fischer and M. Nisbeth Jensen, eds. *Translation and the Reconfiguration of Power Relations*. Vienna: LIT. pp. 21–34.

Holleman, J., 2012. "Rare baseball card tops $1 million in auction." *St Louis Post-Dispatch*, 20 April. www.stltoday.com/lifestyles/columns/joe-holleman/rare-baseball-card-tops-million-in-auction/article_ce495b48-8996-11e1–98ed-0019bb30f31a.html [accessed 7 April 2014].

Korchilov, I., 1997. *Translating History: Thirty Years on the Front Lines of Diplomacy with a Top Russian Interpreter*. New York: Scribner.

LCCC, 2014. *The London Cigarette Card Co. Ltd.* www.londoncigcard.co.uk [accessed 7 April 2014].

Mémorial de l'Armistice, 2014. *Délégations alliée et allemande lors de la signature de l'Armistice de 1918* [*Allied and German Delegations after Signing the Armistice of 1918*]. www.musee-armistice-14-18.fr/#/la-delegation-1918/3160905 [accessed 7 April 2014].

Photographie Provost, 2014. *La guerre 1914 à Toulouse: Galerie de cartes postales anciennes de Toulouse* [*The 1914 War in Toulouse: Gallery of Old Picture Postcards of Toulouse*], CPArama.com. www.cparama.com/forum/la-guerre-1914-a-toulouse-t14489.html [accessed 7 April 2014].

PSA Card Facts, 2014. *Boxing Cards: 1915 Cope Bros. & Company*. www.psacard.com/cardfacts/Hierarchy.aspx?c=32371 [accessed 7 April 2014].

Torikai, K., 2009. *Voices of the Invisible Presence: Diplomatic Interpreters in Post-World War II Japan*. Amsterdam: John Benjamins.

Towersey, L., 2012. *The Card Collecting Home on the Internet*. www.cigarettecards.co.uk/ [accessed 7 April 2014].

Voices of the Armistice, 2008. *An Account of the Negotiations of the Terms of the Armistice Which Took Place Between 8 and 11 November 1918*. Written by the interpreter, Commander W. T. Bagot. Original document catalogue reference: ADM 1/8546/319. http://media.nationalarchives.gov.uk/index.php/voices-of-the-armistice-peace-negotiations/ [accessed 7 April 2014].

8

POWER RELATIONS IN POSTCARDS OF FRENCH FIRST WORLD WAR MILITARY INTERPRETERS

Franziska Heimburger

> If anyone dared, while talking about them, consider the interpreters to be shirkers, all that needed to be said was: Go and see for yourself! … Naturally, over the course of the conflict the losses among interpreters were higher, but of the same order as that of other services. But the real proportion of those killed and wounded compared to those who began service is similar to the losses of the Artillery. This means the survivors have every right to hold their heads high.
>
> (De Saint-André, 1933, p. 7)[1]

In a journal article published in 1933, D. de Saint-André, a former liaison officer in the First World War, mounts a defence against the accusation of interpreters being shirkers by comparing their deaths in service to those of other army units. This brings us right to the heart of the very difficult position of these men – French soldiers or, more rarely, non-commanding officers who were detailed to provide linguistic assistance to the British and later US troops. It is not currently possible to establish a definite figure for the number of men who fulfilled this role. Estimates in 1917 (before the arrival of the American troops) mention a total of 3,000 men. The overall total for the entire war may be between 5,000 and 6,000. They were less directly exposed to the risks of warfare than infantry officers and soldiers and could thus be accused of *embusquage*, shirking their duty (for a general treatment of this question, see Ridel, 2007). Their role of language mediation and general assistance in a coalition situation was not immediately intelligible to their contemporaries, and was far less self-evident than, for example, that of the men sent to serve as German-language interpreters in prisoner-of-war interrogation and on propaganda missions. Furthermore, they lived and worked in areas of the front where they were isolated from the rest of the French army, in contact only with British military personnel and French civilians in the villages behind the front lines.

These untrained individuals had to establish their position in a complex interplay of power relations between different military and civilian French, British and American stakeholders. While power is a concept which has been used to analyse the interaction in interpreting situations (two classic texts are Anderson, 1976/2002 and Tymoczko and Gentzler, 2002; more recently, Fischer and Jensen, 2012), there has so far been relatively little direct use of Michel Foucault's extensive work on power to understand interpreting situations.[2] According to Foucault, power is "a complex strategic situation in a given society social setting" (Foucault, 1980, p. 56), and this essay analyses several photographs for traces of the interpreters' power, understood "not as the traditionally dominating force to monopolize, control, or rule, but as a kind of strategy, disposition, maneuver, tactic, or technique, functioning in a network of relations" (Mason and Ren, 2012, p. 234).

First World War interpreters used different strategies to come to terms with their particular position, often verbalizing their difficulties in letters to their relatives (for a published example, see Cru, 2007). One particular legitimization strategy used photographs taken at the front and posted to the recipients, often close family members. The three pictures from the ICOTI research collection which form the object of this paper must be understood in this context of a private, personal space of communication. This makes them very different from the ones discussed by Michael Kelly in "Staging the Entente in the First World War" (in this volume), which by their nature are public and linked to a larger political discourse, being intended for publication in large-circulation media. This is not the place to look in detail at previous historical work on wartime correspondence and its impact on the combatants in the different armies of the First World War (see, for example, Ulrich, 1997; Anderson and Borlinghaus, 2000; Horne, 2005, p. 905; Watson, 2006). For the present purposes, it is sufficient to underline the crucial role of letters in maintaining links between the home front and the soldiers serving abroad, but also, and this is important for our understanding of these photographs, the extent to which soldiers – within the constraints that they regularly faced – adapted their message to its recipient, depending on their relationship and the intended effect.

My work on language policy and practice in the Allied coalition[3] has led me to investigate the French archival holdings and their British and US equivalents: for the French case, the military archives in Vincennes (hereafter "SHD Vincennes DAT"); for the British case, the National Archives in Kew and especially the WO series for the War Office; and for the United States, the National Archives records held at NARA II in Maryland. These give insights into the French army's logistical efforts to provide the necessary interpreting staff for its coalition partners, and the reception of these efforts among the British and later also US military bodies. Together, the archival material enables a larger contextualization and thus additional depth of analysis.

In the following paper, I submit three postcards to a close reading both as images in their own right and as documents indicative of historical context, with the aim of showing how power relations in a Foucauldian sense are a meaningful tool for understanding the role of a military interpreter.

Complex and ambivalent power inscribed on bodies through military uniform

Full-length portraits of soldiers in uniform, often taken immediately before their departure for mobilization and conserved, in many cases until the present day, by their family and loved ones were a standard trope during the First World War. It appears that in this picture the photographer has attempted to recreate the typical studio setting by making the soldier pose in front of a white wall (Figure 8.1). The resulting picture includes both the white wall and the tell-tale signs of an outdoor setting with a brick wall and gravel on the ground.

This curiously hybrid picture is caught between the seemingly neutral studio of the home front and the outdoor world of the actual front. The relaxed stance is typical of studio portraits of soldiers in uniform, but the inclusion of a notebook in the subject's left hand is rather more unusual for a studio photo (though more common in snapshots taken nearer the front). Archival information from the French army holdings (SHD Vincennes DAT17 N 480–82 contain information sheets on all the interpreters attached to the British base in Le Havre throughout

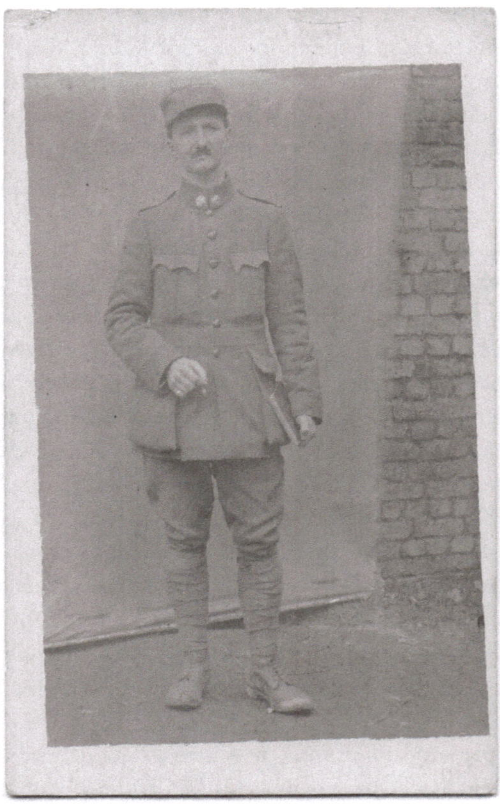

FIGURE 8.1 Postcard sent by Gustave Georges Petit, dated 6 April 1916

the war) suggests that the individual depicted is Gustave Georges Petit, going by his signature, "G. Petit", on the back of the postcard and the available names in the records. He was 34 at the time the picture was taken (the postcard is dated 6 April on verso), and held the rank of a *caporal* – that is, he was a rank-and-file soldier.

The most surprising aspect of this picture for those who would have received it is not immediately intelligible for the casual viewer of our times. The subject is portrayed in the standard uniform of the French interpreters attached to the British troops during the First World War.[4] This was a regular British khaki uniform but with a very distinctive badge depicting a sphinx on either side of the jacket collar and, as a nod to their continued status as French soldiers, a French *képi* on their head. This curious mixture of national uniforms – originally instituted in the first weeks of the conflict when the highly visible French uniform made the French interpreters privileged targets for German snipers as they worked among the khaki-clad British officers and soldiers – became an important element of their collective identity (see Peyrafitte, 1944, p. 1) and a very evocative emblem of their role in coalition liaison. By sending family members a photograph that immediately showed how they could be working for the British army yet still retain their French identity, these military interpreters participated in the legitimization of their particular role.

We can attempt to make sense of the collective identity expressed in this uniform through the Foucauldian understanding of power, which underlines the disciplinary role of power on the body (see Punday, 2000). Indeed, military uniform is an archetypical case of the "normalized body" (p. 512). Like all soldiers, military interpreters experience the power of the military institution as it inscribes its authority on their bodies by imposing a uniform. The hybrid Franco-British uniform of First World War French interpreters thus shows very tangibly their physical inscription into two different regimes of power.

Knowledge as a factor in power relations?

Foucault also writes that "there is no power relation without the correlative constitution of a field of knowledge, nor any knowledge that does not presuppose and constitute at the same time power relations" (Foucault, 1977, p. 27). Ian Mason and Wen Ren choose a slightly different formulation: "Although interpreters often lack institutional power, they may be equipped with power within the exchange as a result of their bilingual and bicultural expertise" (2012, p. 234). What appears at first sight to be a purely hierarchical and social imbalance of power turns out to be based on the interpreter's knowledge, which gives him power and authority. The next photograph (Figure 8.2) reveals precisely this power relationship based on knowledge disparity.

If the full-length portrait of the soldier in uniform was already a very conventional type of representation, the figure of the mounted military leader or ruler has an even longer pedigree. The low-angle shot of this undated photograph corresponds to the monumental character of such paintings and sculptures. The background of

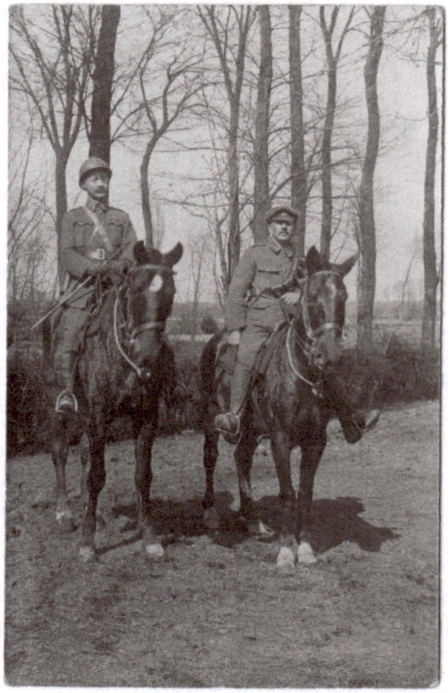

FIGURE 8.2 Undated postcard, sent by Albert Jacques Veillet-Levallée

trees and shrubs seems deliberately free of traces of human habitation or intervention, but also – and rather curiously – completely devoid of all traces of warfare. It seems more appropriate, for example, for a picture of participants in a hunt than for soldiers on active duty.

The placement of the two men induces a clear hierarchical distinction, with the one on the left placed much higher, appearing clearly superior – visual reminders of such pairings as Don Quixote and Sancho Panza are present. The uniforms offer further clues to the relationship between the individuals. The soldier on the left is a French interpreter, again wearing British uniform except for the sphinx emblems on his collar and the French headgear, this time a helmet of the *Adrien* model, which was introduced in 1915 and 1916 for the French troops. The soldier on the right is wearing standard-issue British uniform for rank soldiers and is seated on a smaller horse as well as being on lower ground. All indications are that the British soldier is the French interpreter's batman or groom. British officers were entitled to such a personal servant, whose duties mainly revolved around care of their officer's mount. As interpreters were mounted in order to accomplish their tasks in liaison and in logistical support, they too were entitled to a personal servant, despite usually being rank-and-file soldiers in the French army, or at the very most non-commissioned officers.

The interpreter in this picture can be identified from his signature on the back of the card. He was Albert Jacques Veillet-Levallée, born in 1885 near the French Atlantic coast in Mortagne sur Gironde (SHD Vincennes DAT 17 N 482, interpreter Veillet-Levallée's file). He saw military service from 1904 onwards and was mobilized at the beginning of the war, with the rank of *sergent*, in the French 412th Infantry Regiment. Before the war he had worked as an English teacher at the high school in "Guérét" [sic]; he spoke good English, possibly because his mother, Louise Addison, came from an English-speaking country. He had the equivalent of an undergraduate degree in English from the Sorbonne and had also spent four years living in England. Veillet-Levallée did not become an interpreter until May 1916, presumably serving in his infantry regiment for the first two years of the war.

Combining our analysis of the picture in itself with what we know about the interpreter's life gives additional depth to our reading of the relationship between the two men. Veillet-Levallée was a rank-and-file soldier in the French army, and would certainly not have had the services of a batman during the first years of the war. Furthermore, his social position as a schoolteacher meant he probably would not have had a personal servant or butler before the war. Choosing to have himself portrayed in the manner of a mounted portrait, with his servant at his side, seems to be an attempt to give importance to his role as an interpreter. It is unclear what the relationship between Veillet-Levallée and the intended recipient of the post-card, Sybil Swindeles, might have been – possibly a family member or an acquaintance from his time spent abroad. It seems certain, however, that the interpreter in this case chose to underline the officer-like status he enjoyed and the social distinction which came with his new role. This interpreter gained the power associated with social standing through his specific knowledge.

Establishing and maintaining power relations between language actors in a larger professional setting

According to some Foucauldian analysts, there is a fourth face of power, which is "co-extensive with the social body" (Digeser 1992, p. 982). It is less connected to a specific agent, and therefore detaches "the idea of power from … concerns over responsibility and harm" (p. 983). It is based on shared norms and values, a common understanding of social context. This more diffuse idea of power relates particularly well to professional settings. French First World War military interpreters had to establish themselves in a new role and its associated power network. They worked with colleagues, but also in hierarchical relationships defined by military rank. These relationships were formed through shared experience, but also in large part were based on prior experiences of language and cultural mediation and of social standing. The potential for destabilization presented by these interpreters' situation is well illustrated by the memoirs of another interpreter, who describes his arrival at the *dépôt* as follows:

> The uniforms of the potential interpreters were varied: we all wore that of our respective regiments. The interpreters were equally heterogenous; men had been sent from all parts of France, and I had never realized before what a

> composite country it was, for when you travel from province to province you only see one lot of people, only hear one accent at the time. …
>
> There were men from Boulogne, Calais and the surrounding country – it was said that they had been put down as possible interpreters because they lived near the English coast – some who had just returned from America and said "Sure" instead of "Yes"; merchants, diplomats, journalists, painters, waiters, professional interpreters and a certain number who, no doubt, spoke English tolerably well by the end of the war.
>
> *(Boulestin, 1936, p. 178)*

In the related photographs, interpreters were not only depicted as individuals with their surrounding power network, but also as part of a professional setting. While the first two pictures were relatively straightforward in both their message and their choice of representational modes, the third picture (Figure 8.3) introduces a number of unusual elements. It is worth considering the choice of angle and the choice of framing.

This photograph shows a room with men eating at a table which stretches across the entire diagonal of the picture and seems to extend beyond the frame of the picture. The angle of the photograph indicates that it was taken from a raised platform, possibly a balcony overlooking one end of the large room. This may have been an attempt to fit in as many individuals as possible, but it also condenses the group and underlines its coherence, already indicated by the shared uniforms. At the top end, several individuals are seated at "high table", at right angles to the long row of soldiers. Their uniform indicates officer rank and their position

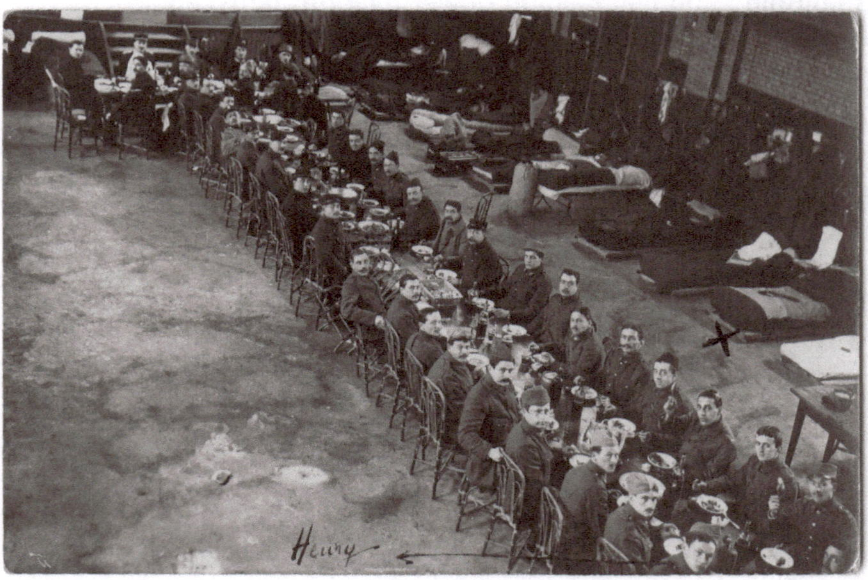

FIGURE 8.3 Meal at the *dépôt* for interpreters at Le Havre, 14 February 1915

suggests overall authority. The whole photograph is an unusual group picture, as the men are photographed during their meal, but on the other hand it is not a snapshot either, as a great number of the men appear to be looking at the camera.

The inscriptions on the back of the postcard confirm the inferences to be drawn from the archival traces of the functioning of the interpreter service on the Western Front. It contains the phrase "campement des interprètes", which explains why this card was put into commercial circulation with the keyword "interpreter". The picture shows the main hall of the *dépôt* for interpreters in Le Havre, where newly selected French soldiers arrived and were held until they were attached as interpreters to a British unit arriving in the port by boat. According to archival records, on 12 February 1915, two days before the card was sent, there were 58 interpreters awaiting attachment to British units in Le Havre (SHD Vincennes DAT 17 N 449, Officier de liaison Ferrere to Colonel Reboul, MMF, 12 February 1915). This would confirm the impression that the table must extend beyond the visible frame of the photograph.

The postcard was sent by an interpreter, Henry Wiesser, who marked both his sleeping mat and his place at the table (indicated on the verso). His individual file (SHD Vincennes DAT 17 N 482) shows that he officially became an interpreter on 15 February 1915, the day after the card was written. This picture must therefore be seen in the context of a young man (born in 1894, he was considerably younger than the average age of French interpreters) anxious to portray the new environment in which he has arrived, and more specifically the new group of which he is now part. It is unclear what, if any, active service Henry Wiesser had seen prior to being detailed as an interpreter. He was due for military service in 1914, but his status as a student of law at the time may have enabled him to delay initial call-up, and further time would have been spent training with his unit before an eventual departure to the front. His choice to send a picture where he is clearly part of a larger whole, a military unit in its own right, seems to indicate a desire for inclusion in the national war effort. At the same time, it shows his professional environment, composed of other men who have just arrived and now have to establish working rapports and cooperate with their hierarchical superiors, seated at the head of the table. The sense of belonging to a larger group with a collective identity, expressed through shared uniform and lived through similar experience before and during the war, is very clear.

Interpreters in general, and military interpreters in particular, are part of a larger professional setting which brings with it hierarchical pressure and power relations. In this case, individual power relations become complex networks under the rule of the Foucauldian "fourth face" of power. This picture enables us to move beyond the face-to-face, binary power relation of the actual interpreting situation and contextualize it in a professional setting. Previous experience, knowledge, social standing and other aspects of power contribute to the relative positioning of the interpreters within their professional group. This is an important part of their experience of their role as a language mediator, and naturally has a significant impact on the performance of their linguistic tasks.

Conclusion

This analysis has only been able to touch on the most salient features of the three photographs, which share a context of private communication about a role in wartime France that was unusual compared to those of the vast majority of the senders' compatriots in the different army units. In subtly different ways, they stress the military nature of their tasks as interpreters, the social distinction acquired in a role where rank soldiers take on responsibilities usually shouldered by officers, and the collective identity which arose among interpreters, reassuring them in a social and political context where their efforts were not necessarily valued. We might rephrase this, taking up the introductory quotation, as different ways of "holding their heads high" as military interpreters. The three postcards have shown situations that reveal the variety of the forms of power exercised over military interpreters, the power exercised by them over others and, finally, the network of power relations within which they act.

Notes

1 My translation. The original French reads: "si quelqu'un avait osé, en parlant d'eux, traiter les Interprètes d'embusqués, il aurait suffi de répondre: 'Allez-y voir!' ... Evidemment, dans l'ensemble, la proportion des pertes des Interprètes a été supérieure, mais sensiblement du même ordre que celle des pertes des autres Services. Mais la proportion réelle, celle des tués et blessés, avec le nombre des hommes engagé, a été de l'ordre des pertes de l'artillerie. Cela donne aux survivants le droit de porter la tête haute".
2 Two significant counter-examples are Lee (2012) and Mason and Ren (2012); Wolf (1997) uses Foucault extensively, but refers more generally to translation and its disciplinary standing.
3 Ongoing PhD project in history entitled "Mésentente cordiale – Language in the Allied coalition on the Western Front during the First World War", EHESS Paris, supervised by Christophe Prochasson (EHESS) and John Horne (Trinity College Dublin).
4 SHD Vincennes DAT 17 N 394 gives a general outline on uniform, and SHD Vincennes DAT 7 N 2288 provides an updated version from 1918 onwards.

References

Anderson, H.-J. and Borlinghaus, H., 2000. *Die deutsche Feldpost im Ersten Weltkrieg 1914–1918: Handbuch und Katalog* [*The German Feldpost in World War I, 1914–1918. Handbook and Catalogue*]. Münster: Geschäftsstelle der AG Dt. Feldpost 1914–18.

Anderson, R. B. W., 1976/2002. "Perspectives on the role of interpreter." In: F. Pöchhacker and M. Shlesinger, eds. 2000. *The Interpreting Studies Reader*. London: Routledge. pp. 209–17.

Boulestin, X. M., 1936. *Myself, My Two Countries*. London: Cassell.

Cru, J. N., 2007. *Lettres du front et d'Amérique* [*Letters From the Front and From America*]. Aix-en-Provence: Publications de l'Université de Provence.

De Saint-André, D., 1933. "L'interprète de liaison" ["The liaison interpreter"]. *Interprètes Militaires Interalliés*, January [s.p.].

Digeser, P., 1992. "The fourth face of power." *The Journal of Politics*, 54(4), pp. 977–1007.

Fischer, B. and Nisbeth Jensen, M. eds., 2012. *Translation and the Reconfiguration of Power Relations*. Vienna: LIT.

Foucault, M., 1977. *Discipline and Punish: The Birth of the Prison*. Translated by A. Sheridan. New York: Vintage Books.

——, 1980. *Power, Knowledge: Selected Interviews and Other Writings, 1972–1977.* Translated by C. Gordon, L. Marshall, and J. Mepham. New York: Pantheon Books.

Horne, J., 2005. "Entre expérience et mémoire: les soldats français de la Grande Guerre" ["Between experience and memory: French soldiers of the Great War"]. *Annales. Histoire, Sciences Sociales*, 60(5), pp. 903–19.

Lee, J.-S., 2012. "Cultural modulation in Chinese into Korean translation and power discourse." *Journal of the Interpretation & Translation Research Institute*, 16(2), pp. 161–77.

Mason, I. and Ren, W., 2012. "Power in face-to-face interpreting events." *Translation and Interpreting Studies*, 7(2), pp. 234–53.

Peyrafitte, J., 1944. *Un du 12e chez nos allies* [*One of the 12th Among our Allies*]. Tarbes: Les Editions A. Hunault.

Punday, D., 2000. "Foucault's body tropes." *New Literary History*, 31(3), pp. 509–28.

Ridel, C., 2007. *Les embusqués* [*The Shirkers*]. Paris: Armand Colin.

Tymoczko, M. and Gentzler, E. eds., 2002. *Translation and Power.* Amherst: University of Massachusetts Press.

Ulrich, B., 1997. *Die Augenzeugen: deutsche Feldpostbriefe in Kriegs-und Nachkriegszeit 1914–1933* [*Eyewitnesses: German Feldpost Letters in the War and Postwar Era, 1914–1933*]. Essen: Klartext.

Watson, A., 2006. "Self-deception and survival: mental coping strategies on the Western Front, 1914–18." *Journal of Contemporary History*, 41(2), pp. 247–68.

Wolf, M., 1997. "Translation as a process of power: aspects of cultural anthropology in translation." In: M. Snell-Hornby, Z. Jettmarová and K. Kaindl, eds. *Translation as Intercultural Communication.* Amsterdam: John Benjamins. pp. 123–33.

9

STAGING PRISONERS OF WAR

The interpreter figure in First World War postcards

Michaela Wolf

During the First World War, an estimated six to eight million men were held in camps as prisoners of war (POWs) – yet historians have only recently begun to examine this captivity experience seriously (see Jones, 2011). The pertinent research touches on various dimensions, including military questions, issues around foreign policy and nationality, and economic aspects in terms of the deployment of POWs as forced labour (Oltmer, 2006, p. 9). The First World War entailed an unprecedented degree of contact with people from other nationalities, among both the civilian population and the military. These encounters brought with them a need for interpreters, especially in the context of prisoner interrogation by officers and the presence of different nationalities in most camps. Postcards issued mainly for propaganda purposes often "staged" those interpreters in their motifs, not least as an attempt to prove that the prisoners were being treated humanely.

The postcards investigated in this paper show Russian and British prisoners of war in German captivity.[1] Examining them enables an exploration of the nature of the dependencies and power relations between the various agents involved, and the ways in which the staging of interpreter figures in the pictures contributes to the enactment of the relationship between officers and prisoners. On the basis of these and similar questions, I analyse the interpreting scenes with the help of Erving Goffman's dramaturgical theory.

Postcards as a medium of war propaganda

At the time when the First World War broke out, postcards were immensely popular and formed the economic basis of a vast and diverse printing industry throughout Western Europe. The First World War was a "total war" (Jones, 2011, p. 375) for its European participants, and that totality was largely due to the creation of modern propaganda; this was reflected in the mass of postcards circulating during the war. The military or "field post" was free of charge, and the postcard medium burgeoned as the war progressed. Indeed, the First World War has sometimes been called a "war of the pictures" (May, 2004, p. 144).

In times of war, governments tend to have an interest in controlling the exchange of communication through censorship and propaganda. Thus, the function of the postcard was soon not simply to send snapshots of a world at war, but to direct the minds and motivations of the people involved: postcards were intended to mobilize the population, downplay military losses and consolidate the public's belief in its own nation's supremacy and hopes of a splendid victory (p. 144). In such a context, pictures of prisoners of war offered particularly profitable potential for political instrumentalization (Becker, 2006, p. 26).

Interpreters in the First World War

Military interpreters on both sides were mostly soldiers, officers or civil servants; as an exception, though more and more often with the war's expansion, individual civilians were recruited from the local population (Kraus, 2009, p. 790; see also Heimburger, 2012). Their areas of responsibility were very broad. Those employed away from the front lines provided for communication among the Allied troops, while contact with local institutions was particularly important in occupied zones. Others worked with deserters and prisoners, mostly interpreting the interrogations and monitoring prisoners' mail. In many instances, military interpreters analysed and translated documents that had been captured, additionally collecting and preparing all information available on the situation of the enemy in the relevant sector of the front before handing it over to the army headquarters. Towards the end of the war, military interpreters also conducted negotiations with the enemy and were often given a surprisingly free hand in the process (Wolf, 2012, p. 175). With reference to work in POW camps, a Prussian decree of July 1915 states that one interpreter should be responsible for about 1,000 to 1,500 prisoners (Kraus, 2009, p. 1061).

Interpreting activities varied among the different combatants in the First World War, primarily according to the demographics of the nationalities involved. In the Habsburg Army, interpreters were coordinated by the "Evidenzbureau", a unit of the War Ministry (see Wolf, 2012, p. 174). This office mainly recruited men with a multilingual background and focused on their skill in understanding different dialects. To give a few examples, 42 interpreters and intelligence officers were successfully deployed during the Austrian recapture of Galicia in 1915, and on the Italian front in April 1918 there were 220 officers and 1,000 ordinary soldiers working as interpreters (Fuchs, 1994, p. 96). The interpreters were trained at interpreting schools distributed all over the central parts of the Habsburg Monarchy (Ronge, 1930, p. 20, p. 273). In summer 1917 a further school was established in Lviv, Galicia, in order to remedy the persistent lack of Russian-speaking interpreters on the Eastern Front. Interpreting from Russian often created major problems because, inside the Habsburg Monarchy, Russian was mostly spoken by

Ruthenians, who were often regarded as Russophile and thus deemed unreliable (Pethö, 1998, p. 168, p. 368).

Prisoners of war as protagonists in the war drama[2]

In his dramaturgical theory, the US sociologist Erving Goffman presented social life as a staged drama. In this drama, each individual performs a certain role,[3] which varies according to his or her audience and reflects his or her societal relationships (Goffman, 1961, p. 87). Goffman did not develop a theoretical approach aiming to explain all the elements of the social world, but rather a microsociological analysis of the face-to-face activities of the actors participating in everyday life situations. My analysis of the pictures "staging" prisoners of war in German captivity draws on Goffman's social theory, with a particular focus on the social interactions that can be detected in these interpreting situations. It addresses questions which may enhance our understanding of the conditions underlying the construction of the interpreter figure through visual documentation. These questions include: Does the image composition provide clues as to the social and hierarchical position of the interpreter within the communicative system of the POW camp? Can we detect any signs of superiority or inferiority in the situation of interpersonal exchange? What are the discursive connections between the pictures and the accompanying texts, and what do these connections tell us about the enactment of the power relationships? What are the frontiers, spaces or margins created by the interpreter figure?

The first photo (Figure 9.1) features Russian prisoners being interrogated by German officers and an interpreter.[4] According to the caption, the location is

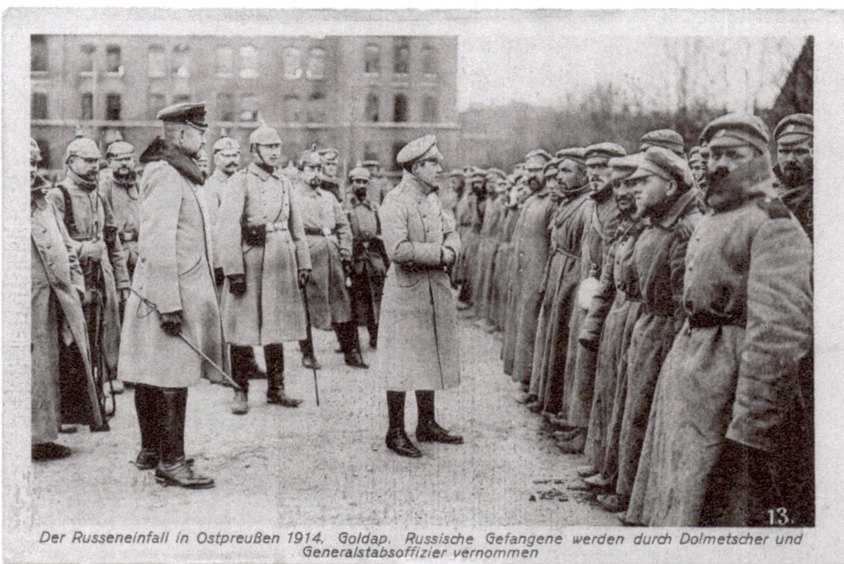

FIGURE 9.1 The Russian invasion of Eastern Prussia, 1914

Gołdap, a small Polish town that today lies on the border with Lithuania. During the First World War it was the scene of fierce fighting on the Eastern Front, which passed through the town twice. After the outbreak of war, the Russian westward movement was rapid, but quite inefficient due to communication and logistical problems within the Imperial Russian Army. A first Russian push into East Prussia was promptly followed by the devastation of the Second Russian Army at the Battle of Tannenberg (Neiberg and Jordan, 2012, pp. 32–33). It may be assumed that the picture was taken in the wake of this battle. The back of the postcard tells us that the photo was taken by the "Court photographer A. Kühlewind, official war photographer in Königsberg" (today's Kaliningrad), a fact that supports the view of the postcard as a propaganda medium intended to persuade the German civilian population of their Army's supremacy and its masterly handling of situations like the one presented on the postcard. In the first years of the war, the Hague Convention of 1899, which established principles for the treatment of prisoners of war, was still widely observed; however, this changed dramatically towards the end of the war, with a deterioration of conditions and increasing malnutrition or starvation and violence (Becker, 2006, p. 28; Jones, 2011, p. 372).

According to Goffman's dramaturgical approach (1961, pp. 41–67), each individual plays a role, in an intentional and usually manipulative way, with the aim of managing others' impressions as part of a "team". In the context of this postcard, the institutionalized interpreting activity takes place on what Goffman calls the "front stage" – the zone where the "ensemble" meets the audience. That stage is constituted by the long row of Russian prisoners, who appear to be frightened, yet also curious about the role apparently being imposed on them through the photographing process; however, the German officers opposite the prisoners are also part of the front stage. They convey a complacent impression, obviously satisfied with their early military success and perhaps the bearers of "entrusted secrets" (see Goffman, 1971, p. 143) – secrets they are obliged to keep in order to maintain the team's integrity; in this case, the secret of the prisoners' future fate. Especially in the first months of the war, interrogations in a line-up like this were quite common. Psychologically, soon after being captured the soldiers would be less prepared to keep their "strategic secrets" – secrets pertaining to the intentions and capacities of a team which it conceals from its audience – and likely to deliver more information about combat situations than during classical interrogations. Consequently, most regiments had their own interpreters, who often accompanied the troops during military encounters, both as civilians and as members of the military.

The picture poses the interpreter in the centre, between the two opposing groups. He is a young officer who holds his hands crossed, a gesture which might signal insecurity and lack of experience, but could also be an attempt to intimate reluctance in order to avoid a deeper involvement in the scene; or perhaps it is simply a cold day. Yet this interpreter certainly knows that he must maintain the "façade", which Goffman sees as a repertoire of means of expression that underlies any interaction. When an actor opts for a certain social role, he or she will be aware that for that role a certain façade already exists and must be fulfilled. In this picture, the interpreter's social role is twofold: as well as interpreting the officer's questions, he is keeping apart the two groups – the German officers with their tidy uniforms and self-assured gestures, and the rather poorly dressed Russian prisoners, characterized by concern and fear. The first group figures Western, European civilization, the latter Slavdom and Asianness. The interpreter is positioned at the fault line of the staged drama, in an in-between space. As a "go-between", the interpreter understands both sides and hypothetically has access to each side's knowledge. Goffman tells us that the activity of the go-between, as an individual, is bizarre, untenable and undignified, vacillating from one set of loyalties to another. Here, however, the picture outwardly suggests that the young officer's concern is limited to his role as mediator for his own people, who have certain expectations of his performance and physically back him in his endeavour to satisfy his task. From the audience's point of view, the vanishing lines locate the interpreter at the very key position of the photo, making him the person upon whom all the picture's tension converges. The photographer thus positions himself in such a way as to focus the entire responsibility for the scene's discourse on the interpreter. The group of soldiers on the left are, in Goffman's terms, "back stage": they are not openly participating in the scene, but are holders of entrusted secrets, which they are obliged to keep because of their relationship with the team to which the secret refers.

Russian prisoners in German captivity are also the subject of the second postcard (Figure 9.2). The audience of this picture can detect three different stages: the front stage, featuring the interpreting activity; one back stage in the picture's foreground, presenting the Russian prisoners; and the second back stage, in the background of

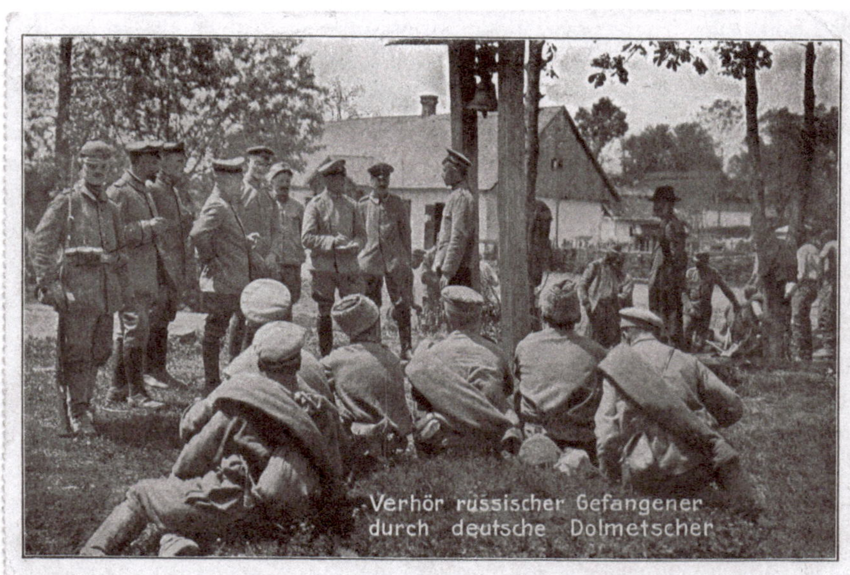

FIGURE 9.2 Interrogation of Russian prisoners by German interpreters

the photo, with peasants. The figures on both back stages are attentively watching the proceedings on the front stage; both back stages suggest a sort of refuge, a locus for restoring the façade necessary for the performance, and a "site of confidentiality" (Raab, 2008, p. 71). The interpreter, holding a writing pad, appears fully integrated into the group of German officers, some of whom may also be interpreters.[5] As such, he does not create a dividing line between the various stages as is the case in other pictures. The officers surrounding him also adopt a rather relaxed attitude. With the exception of two officers who are looking into the camera and to some extent leaving the ensemble by connecting with the audience, all are listening keenly to the interrogation and are united by their gaze at the Russian officer under interrogation. This officer's central position in the picture additionally gives him the function of connecting the various stages. The Russian prisoners, too, are watching the scene. They are apparently waiting to be interrogated and, being seated, allow the audience to see the interpreter-officers. The scene is completed by the back stage of the peasants. They underline Goffman's "presentation of self in everyday life": the audience sees local people who make the interrogation scene an ordinary one. Their possible involvement in the situation may be signalled by the presence of an officer on the back stage, perhaps a liaison officer in charge of establishing contact with the local population. The farmer's house suggests a certain wealth, and the thatch roof behind the bigger house intimates that the location is somewhere in Eastern Europe. The postcard was sent by "cannoneer Emil Wegerich" to "Frau Anna Wegerich" in July 1916. Very likely the picture was not produced in the first phases of the war, as the behaviour of the participants implies some experience of the interrogation situation. The professional demeanour of the interpreter (or interpreter group) and the back stages encircling the interrogation scene testify to a masterly performance, which is an indispensable element of the postcard's propaganda function. Goffman contends that in order to guarantee smooth communication, all individuals involved in an interaction take pains to fulfil the expectations assigned to their social role (1961, p. 163). In such a way, the normative expectations of the group and the actual conduct of the individual come to fit a nexus which reinforces the model of social roles in question. This is well illustrated by the present postcard: the actors on the various stages clearly belong to their specific setting, yet they all behave according to functions assigned to them by the interrogation situation more generally.

The third picture (Figure 9.3) represents a clear shift in the staging of prisoners of war. It is a group photo of British prisoners, harmoniously lined up and all looking towards the audience into the camera. The caption "Kriegsgefangene Engländer mit Dolmetscher"[6] is misleading, as the picture also shows French soldiers, including some from the Maghreb. The postcard was sent by Anton Erhardt to his brother in Ellwangen, Württemberg, in August 1915. It is not a typical portrayal of POWs, as there is only one level of representation, and therefore only one stage: the front stage. Some elements – such as the laundry hung out on the right edge of the picture – indicate that the photo was taken inside the prisoner camp. The interpreter is difficult to identify; he might be the figure seated in the first row, in

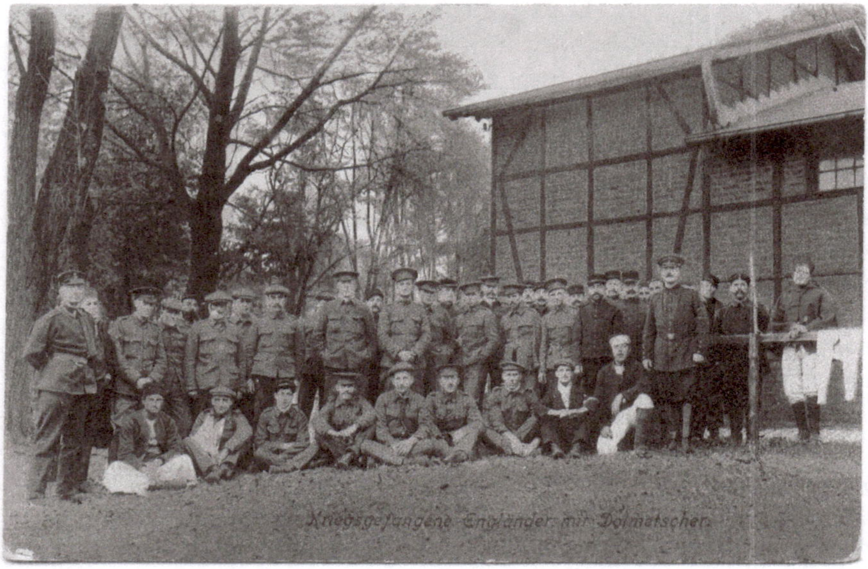

FIGURE 9.3 English prisoners of war with interpreter

civilian dress, next to an injured Maghrebian soldier. However, the camouflaged position of the interpreter in the picture is counterbalanced by the caption, which clearly indicates that an interpreter is part of the scene. The situation is relaxed – the prisoners look serious but not frightened, and some are even smiling. The propaganda function of the postcard is once again evident, and delivers two main messages. First, the signal to the German audience is one of success and pride, and the assurance that the German troops are acting in a civilized manner. Second, by evoking a school photo, the picture enacts a positive scene for the opposing party as well, giving evidence of the fair treatment of the prisoners by the capturing forces and showing the good conditions that the prisoners enjoy in the camp. This message counteracts the Entente propaganda that repeatedly denounced massive violations of the Hague Convention by the Central Powers. The key reference to the interpreter also stands for the richness of human resources at the German Army's disposal, which are bound to help it win the war. The importance Goffman attributes to the ensemble well illustrates the function of this photo: in his interactionist theory, the ensemble strives for cooperation in order to outwardly show that the different members are playing a common social role (Goffman, 1961, p. 96); in the present case, all actors staged in the photo give the impression of jointly creating and maintaining a certain situation, perhaps by distorting real facts or by concealing information.

Concluding remarks

In all three pictures, the captions draw attention to the figure of the interpreter and ascribe the major role in the photo to him, although not all of them portray an actual interpreting activity. The captions thus do not merely support the picture, but control it – for, evidently, the focus on the interpreter gave the capturing forces (in these cases, German) a propaganda opportunity to stress their attentive treatment of the prisoners, in order either to counteract negative rumours or to call for better treatment of German prisoners by the enemy (Jones, 2011, p. 374). The interpreter is not explicitly ascribed the role of a go-between, but in the first two pictures is shown as being concerned to mediate yet still strongly rooted in his own troop. The interpreter's secrets, which in Goffman's sense help to control the flow of information between the various social roles, are both entrusted secrets and "inside secrets", which mark an individual as being a member of a group. In all three postcards the interpreter acts on the front stage, which is not surprising in view of the central role attributed to him; there are no signs of a secret understanding with any figure from the back stages. It is the consideration of the various secrets in these photos that can amplify our perspective on the motives for the various actors' performances, not least because the military context in general bears a high potential for secrets. Focusing on such secrets helps to answer the question of which stages appear as self-contained and which as systems intersecting with other systems, the latter being the case in the third picture. The power balance in the first two pictures is quite clear: the interpreter's position is itself a gesture of superiority, and the interpreter is also visibly backed by his social group, which thus inscribes him into the social system of the protagonists of the front stage. Figure 9.3 more explicitly than the others flaunts its own propaganda function, by ostentatiously signalling de-hierarchization through the linear arrangement of all the participants in the scene.

This analysis of the three postcards featuring interpreters in POW camps has shown that the interpreter figure is actively used to compose the pictures. At the same time, visuality as an analytical perspective emphasizes the interpreter's notorious ephemerality. What is true of the profession more generally also, then, applies to the postcards' captions, as the pictures acquire their specific meaning only through the correlation of image and writing – two elements which are constantly reinscribed. It is the caption which temporarily fills the image with a certain form of content – another metaphor for the interpreter's ephemerality.

Notes

1 These postcards were selected from the ICOTI research collection.

2 For their help in discussing these pictures, I would like to thank Wolfram Dornik from the Ludwig Boltzmann Institut für Kriegsfolgenforschung, Graz, and Gerhard P. Groß from the Zentrum für Militärgeschichte und Sozialwissenschaften der Bundeswehr, Potsdam.

3 Today, the concept of "role" has largely been superseded by that of "habitus" as developed by Norbert Elias and later Pierre Bourdieu. In the present paper, "role" is mainly applied on a metaphorical level.

4 This is also the content of the caption: "Der Russeneinfall in Ostpreußen 1914. Goldap [sic]. Russische Gefangene werden durch Dolmetscher und Generalstabsoffizier vernommen" ("The Russian invasion in Eastern Prussia 1914. Goldap. Russian prisoners being interrogated by interpreter and officers of the general staff").

5 The caption signals the presence of a larger number of interpreters: "Interrogation of Russian prisoners by German interpreters" ("Verhör russischer Gefangener durch deutsche Dolmetscher").
6 Most of the postcards were printed in various editions. This one, for instance, may be found today in the collectibles market in a black-and-white version, or with the additional text "auf der Wahner Heide" ("on the Heath of Wahn").

References

Becker, A., 2006. "Paradoxien in der Situation der Kriegsgefangenen 1914–18" ["Paradoxes in the situation of prisoners of war, 1914–18"]. In: J. Oltmer, ed. *Kriegsgefangene im Europa des Ersten Weltkriegs [Prisoners of War in Europe in the First World War]*. Paderborn: Ferdinand Schöningh. pp. 24–31.

Fuchs, M., 1994. *Der österreichische Geheimdienst [The Austrian Secret Service]*. Vienna: Ueberreuter.

Goffman, E., 1961. *Encounters: Two Studies in the Sociology of Interaction*. Indianapolis, IN: Bobbs-Merrill.

——, 1971. *The Presentation of Self in Everyday Life*. Harmondsworth: Penguin.

Heimburger, F., 2012. "Of go-betweens and gatekeepers: considering disciplinary biases in interpreting history through exemplary metaphors. Military interpreters in the Allied coalition during the First World War." In: B. Fischer and M. Nisbeth Jensen, eds. *Translation and the Reconfiguration of Power Relations*. Vienna: LIT. pp. 21–34.

Jones, H., 2011. *Violence Against Prisoners of War in the First World War*. Cambridge: Cambridge University Press.

Kraus, J., 2009. *Die feldgraue Uniformierung des deutschen Heeres 1907–1918 [The Field-grey Uniform of the German Army 1907–1918]*. Vol. 1–2. Vienna: Verlag Militaria.

May, O., 2004. "Die Ansichtskarte als mentalitätsgeschichtliche Quelle" ["The postcard as a source for the history of mentalities"]. In: R. W. Keck, S. Kirk and H. Schröder, eds. *Bildung im Bild. Bilderwelten als Quellen zur Kultur-und Bildungsgeschichte [Education in the Shot. Pictorial Worlds as Sources for the History of Culture and Education]*. Bad Heilbrunn: Klinkhardt. pp. 126–65.

Neiberg, M. S. and Jordan, D., 2012. *The Eastern Front 1914–1920*. London: Amber Books.

Oltmer, J., 2006. "Vorwort" ["Preface"]. In: J. Oltmer, ed. *Kriegsgefangene im Europa des Ersten Weltkriegs [Prisoners of War in Europe in the First World War]*. Paderborn: Ferdinand Schöningh. pp. 9–10.

Pethö, A., 1998. *Agenten für den Doppeladler: Österreich-Ungarns Geheimer Dienst im Weltkrieg [Agents for the Double Eagle: Austria-Hungary's Secret Service in the World War]*. Graz: Stocker.

Raab, J., 2008. *Erving Goffman*. Konstanz: UVK.

Ronge, M., 1930. *Kriegs-und Industriespionage [Espionage of War and Industry]*. Vienna: Wehle & Höfels.

Wolf, M., 2012. *Die vielsprachige Seele Kakaniens: Übersetzen und Dolmetschen in der Habsburgermonarchie 1848–1918 [Kakania's Multilingual Soul: Translating and Interpreting in the Habsburg Monarchy, 1848–1918]*. Vienna: Böhlau.

PART IV

Second World War

Power in many guises

10

FRAMES AND THE INTERPRETER IN THE IMPERIAL WAR MUSEUM PHOTOGRAPHIC ARCHIVE

Hilary Footitt

The Second World War was a watershed moment in the use of cameras to record armed conflict. As the war went on, nearly every Allied fighting unit in the field had its own basic team consisting of a stills photographer and a cine photographer. In addition, a small group of freelancers covered the war directly for papers and magazines and a huge, if indeterminate, number of amateur photographers – civilians and soldiers – used their (now relatively cheap) portable cameras to memorialize war experiences. The final output of all this image-taking was enormous – Jorge Lewinski suggested that, of the millions of negatives held by the Imperial War Museum (IWM), London, in its photographic archive, nearly half were related to the Second World War (1986, p. 95).

Surprisingly, though, historians are often reluctant to accord such a treasure trove of visual documentation much status beyond that of the purely illustrative, providing a visual exemplar for an already chosen analytical theme. A tendency to concentrate on the production side of the photograph – the role of the photographer, the intermediary agencies leading to publication – has made war photographs appear to be potentially biased accounts of particular events, "the photographed event" frozen in time by the photographer to produce an institutional or personal, and hence suspect, testimony. However, as Ariella Azoulay argues (2008; 2012), there is a key distinction between the "event of photography" and the "photographed event". A photograph, she points out, is never the result of a lone individual, but rather of a range of protagonists. The co-presence of other people at the time when the photograph is taken, the relationship between the photographed persons and the spectators, the rearticulation of the details of the photograph by subsequent spectators – all these position photography as a common rather than an individual endeavour: "the photograph always preserves traces of the gaze and action of additional protagonists", it is "made up of an infinite series of encounters" (Azoulay, 2012, p. 54, p. 6). By turning attention away from the "photographed event" to the encounters inscribed in the "event of photography", Azoulay seeks to disrupt the more traditional polarized categories of inside/outside, photographer/photographed person, spectator/subject of observation and to construct instead "a political space where human beings look at one another" (p. 53). Her technique reads photographs through three frames or gazes: what she calls the "orienting gaze" (providing basic information about the photographed people), the "professional gaze" (borrowing from the discourse of art to examine the aesthetics of composition) and finally "the civil gaze", which she claims will allow "all participants populating the civil space of the photograph to meet on the same plane" (p. 121).

These frames, which focus on the event of photography rather than the photographed event and imagine the chain of human encounters which are being inscribed in a photograph, arguably have a particular relevance when visualizing the essentially oral work of interpreters. Interpreting is, after all, fundamentally concerned with encounter, communicative encounter, positing at the very least the existence of two people or groups who are seeking to understand each other. Examining photographs in the frame of Azoulay's civil gaze may help us to see interpreting within a network of human relationships that were formed at the time and are continued today through the gaze of spectators. Adopting Azoulay's perspective, this contribution examines selected Second World War images, catalogued under "interpreter" in the IWM's photographic archive – all, with the exception of Figure 10.1, taken from the IWM digitalized collection. Recent work on languages in war (Footitt and Kelly, 2012) has identified representative situations in which interpreters typically operate in wartime – communicating with clandestine forces, liaising between the army and civilians, and dealing with the aftermath of war (prisoners of war and diplomatic peace negotiations) – and the images which follow have been chosen from the IWM archive in order to explore each of these potential contexts for interpreter activity.

Communicating with clandestine forces

Figure 10.1 comes from the collection deposited in the IWM by Colonel David Smiley, a British SOE (Special Operations Executive) agent and allegedly the model for John le Carré's iconic spy character George Smiley. Three large albums of photographs, presumably taken by Smiley himself and his British companions, show the life of underground fighters in Greece and Albania, assisted by SOE forces. Both figures in this photograph are locally recruited interpreters and, as Smiley's memoirs confirm, their different styles of dress – Halit Kola's Moslem baggy trousers, Shaqir Trimi's shirt and riding breeches – reflect the different ethnicities and political affiliations of the groups which SOE were meeting in Albania: Halit Kola was "brother to a Gheg chieftain called Bilal Kola, whose allegiance was to Kupi", Shaqir Trimi was "a schoolmaster from Shëngjergj, a Zogist in politics" (Smiley, 1984, p. 114). In a sense, the interpreters are clothed with their particular ethnic identities which are crucial to the linguistic roles they perform with Smiley and his British colleagues.

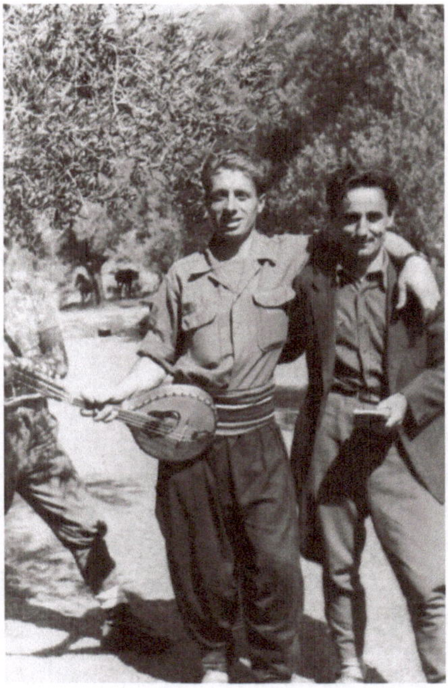

FIGURE 10.1 Second mission: the two interpreters Halit Kola and Shaqir Trini at Mukaj, August 1944

Rather than being positioned functionally as interpreters working with SOE officers, these men are imaged as individuals, looking out at the photographer in a relaxed pose, as if resting or on holiday, snapped informally with someone passing by caught in the same shot. Instead of weapons, one of them holds a musical instrument, suggesting a different world of peace and relaxation. According to Smiley, Halit Kola "had a good voice, and owned a guitar, which accompanied him everywhere, and he often sang for us" (p. 114). The photograph is taken outside, probably in one of the forest camps where the local resistance armies, supported by SOE, would be hiding. What is evident from the image is that the men concerned have a companionable relationship with the photographer: they smile and look over at him with confidence.

Other protagonists would also have been present in this scene – SOE agents watching Smiley posing the interpreters, and local soldiers going about their business in the camp. Smiley's image suggests a bucolic moment of peace and recreation within the dangers and uncertainties of a guerrilla war, fraught with the political and ethnic differences signalled by the interpreters' styles of clothing. As local interpreters, these men would be left behind when the SOE agents were ordered to leave. Smiley's account makes clear his "consternation and dismay" (p. 151) on hearing that the Allied authorities had refused his request to evacuate the interpreters with them when they left Albania. Halit Kola would escape to the mountains to fight the Russian-backed partisans and die in the process, whilst

Shaqir Trimi would be captured by these same partisans, tortured, and sentenced to 15 years' hard labour for having helped the British. In many ways, Figure 10.1 testifies to the price which local interpreters paid, and continue to pay, for their association with incoming foreign forces.

Liaising with civilians

The setting of this photograph is an airbase leased by the British from the Portuguese at Lajes on the Azorean island of Terceira, although the IWM description calls the airbase "Lagens", presumably in imitation of the Portuguese pronunciation (Figure 10.2). The image is captioned as "An RAF (Royal Air Force) Sergeant Interpreter discusses domestic business with Portuguese women who have undertaken laundry work for RAF personnel in the tented encampment at Lagens".

Unlike the local interpreters in the Smiley photograph (Figure 10.1), the interpreter is British, an integrated member of the Allied forces, in uniform, with an armband and a caption designation which places his position as interpreter on a clear military footing. The distinction between civilians and military is visualized in the clothes they are wearing – the RAF uniforms of the men, the headscarves and dresses of the women – and indeed in the gender roles being depicted: women with laundry, men outside a "Guardroom". The physical centre of the photograph, and the pivot of the action, is the Sergeant Interpreter, who stands considerably taller than those around him. The scene is an amicable one – other RAF personnel

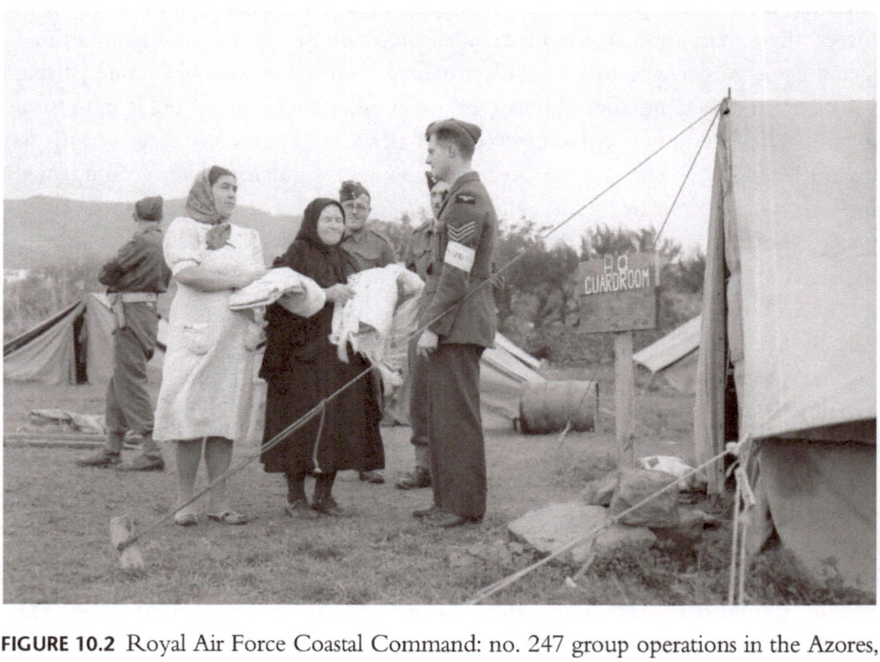

FIGURE 10.2 Royal Air Force Coastal Command: no. 247 group operations in the Azores, 1943–1945

look unconcerned about what is taking place, and the camp itself is noticeably unwarlike.

The image shows a small part of what was a much larger British camp at Lajes. The RAF serviceman looking away into the distance is gazing over what had become in effect British territory for the duration of the war. Around the encounter in the photograph, there is likely to have been a host of other tents and British personnel. Unlike the Smiley image (Figure 10.1), where the interpreters are at home in their own country, the RAF interpreter is on foreign soil that has been visually claimed as British through the notice in English, "Guardroom". This naming of occupied space, what Mary Louise Pratt calls "the power of naming" (2008, p. 32), involves an important emotional dimension, helping those who are outside their normal home environment to feel more comfortable and less foreign when they are abroad. Paradoxically, the foreign interpreter in this image has become the host in territory which is now occupied and appropriated, whereas the Azorean inhabitants find themselves as guests, invited onto the base to perform particular tasks supporting the military. The amicable meeting between Azorean women and RAF interpreter visualizes what Pratt has termed "the contact zone", "the space and time where subjects previously separated by geography and history are co-present, the point at which their trajectories now intersect" (p. 8). The military interpreter chatting sympathetically to local civilians represents the moment at which the typically asymmetrical relationships of war come together, embodied in a figure drawn from the occupying force, the interpreter, tasked with communicating with those who, willingly or unwillingly, are now being occupied. With the more recent "cultural turn" in the Western military (Porter, 2009), where winning hearts and minds – non-kinetic warfare – takes precedence over physical and technological force, the contradictions implicit in the RAF Sergeant Interpreter's encounter with Azoreans in Figure 10.2 have been replicated over and over again in other sites of contemporary conflict like Iraq and Afghanistan.

Dealing with prisoners of war

Figure 10.3 differs markedly from the previous two photographs, taken as it is at the end of the war and in formerly enemy territory. The image is described at some length by the IWM: "Japanese troops leave Bangkok. A British officer of a Gurkha battalion gives the order, through an interpreter, for Japanese prisoners to hand over all sharp objects such as knives, needles and razors before boarding trains that will carry them to Prisoner of War camps outside the city of Bangkok".

The photo is composed with two spaces – an inside and an outside. The disorder, combined with the static waiting position of the Japanese, indicates the powerlessness of those inside the enclosing fence of boxes. All eyes within this space are directed towards the two figures outside the group: the man on the extreme left, dressed as a British officer of the Gurkha battalion and, beside him, with his hand raised, the interpreter. The majority of the photograph, within the enclosed space, is silent. The only words being uttered will come from the two

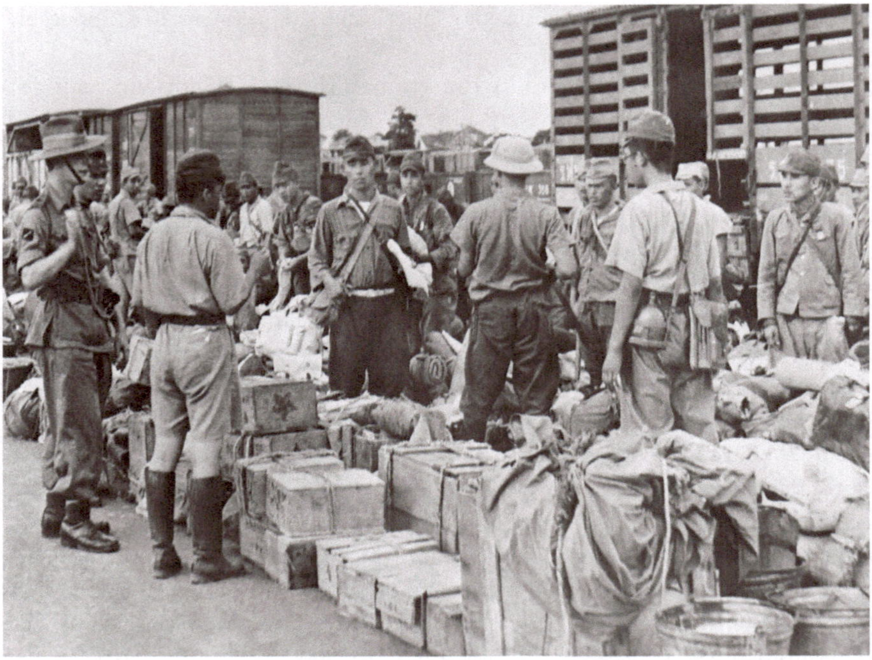

FIGURE 10.3 Japanese troops leave Bangkok, 1945

figures outside. Action is suspended, an order is being given, and the interpreter has a function to perform, that of transmitting the order and hence stimulating movement within the static, closed group. Although the interpreter stands next to the figure of power, he is visually distinct from the British officer, a non-white, wearing a different uniform (cap, short trousers) from the officer. The interpreter evidently plays a secondary role, the mediator of orders rather than the initiator of action. None of the protagonists seems to be aware of the camera. What is important to them, and central to the composition of the image itself, is the task in hand, to initiate movement which will affect the waiting group of prisoners.

Outside the frame of this crowded photograph, there would have been a series of other trains, with more waiting prisoners, and occasional Allied troops overseeing the operation. An estimated 50,000 Japanese soldiers surrendered to the victorious Allies, and the logistics involved in transporting them to POW camps were considerable in all areas, including the one caught in this image. The very presence of captured Japanese troops, and the captioned concern to take sharp objects away from them, references a whole hinterland of problems around the capture of prisoners, in particular the danger of false surrenders when Japanese personnel committed suicide in the act of surrendering (Dower, 2000, p. 87, p. 108).

In this particular encounter, the Allied authorities are represented by members of a British Gurkha regiment. Gorkhas (whom the British called Gurkhas) came from Nepal and were originally recruited into the British imperial armies in the middle of the nineteenth century. By the Second World War, British Gurkha regiments had been organized into a system of separate British and Gurkha military ranks, with Gurkhas never commanding white troops and usually acceding to officer commissions only by moving up through the ranks (Farwell, 1985, p. 132). The different uniforms and ethnicities of the two soldiers image this dual structure of the British regiment and its clear origins in the country's imperial past. In this situation, the interpreter is a subordinate part of the prevailing military hierarchy, situated within the Allied structure but given a secondary role. Communicating with Gurkha troops meant that British-born officers had to be able to speak some of the Nepalese language, Parbatiya, known in the regiments as Gurkhali (p. 115). Viewed from this perspective, Figure 10.3 visualizes a more complex linguistic interchange than "language one to language two", a communicative situation in which the protagonists are engaged in a highly sophisticated language transfer.

Negotiating for peace

Figure 10.4 shows Vladimir N. Pavlov, Stalin's interpreter, together with Stalin and Churchill at the Yalta peace conference in February 1945. Unlike the previous photographs, this image was clearly intended to be diffused as widely and as quickly as possible in the contemporary press around the world. It is a public photograph, depicting important statesmen at a conference which would have major international consequences.

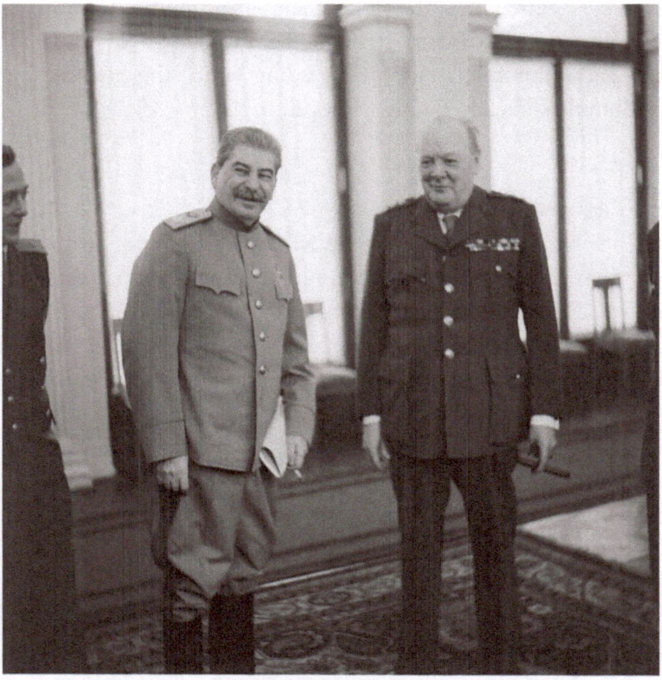

FIGURE 10.4 Yalta Conference, February 1945

For the first time in this essay, the photograph is an interior shot, taken in the Livadia Palace. All the men, including the interpreter, wear formal uniforms, Churchill holding his trademark cigar. The interpreter has been partially cut out of the frame, although the IWM has another version of the image where more of the interpreter's body appears. The IWM description references both the partial invisibility of the interpreter and the nature of the encounter that appears to be taking place: "Winston Churchill shares a joke with Marshall Stalin (with the help of Pavlov, Stalin's interpreter, just visible left) in the conference room at Livadia Palace during the Yalta Conference". The interpretation of jokes is one of the most difficult of professional feats, but the composition is singularly lacking in any sign that actual communication is taking place between the three parties. The men face out rather than in towards each other, and there is little sign of eye contact between them.

All three principals are gazing at what is probably a line of spectators standing in front of them, beside the photographer himself, in a staged encounter intended for public consumption. Both Stalin and Churchill would have had a large entourage of advisers and supporters, and indeed may have recognized particular personalities within this spectator group. Putting two world leaders together, and having them smile for the purposes of a photograph, does not necessitate the physical presence of an interpreter; it is the headline conceit that the two men are getting on well together, evidenced by a shared joke between them, that clearly requires the visible services of a language intermediary. In this photograph, the interpreter is needed solely as a symbolic indication of the hypothetically good relations that exist between the two leaders, rather than as a channel of communication. The partial visibility/invisibility of Pavlov is a strategic demonstration of the good rapport between Stalin and Churchill, a rapport regarding which the invisible spectators of the scene are doubtless already highly sceptical.

Conclusions

This selection of photographs from the IWM has raised issues around the professional role of interpreters in war which are as relevant today as they were in the 1939–45 conflict. The vulnerability of local interpreters who are left behind when intervening forces evacuate; the weaponizing (Rafael, 2009, pp. 1–23) of the military interpreter as a component of counterinsurgency strategies to win hearts and minds; the linguistic complexity of interpreting in tense multi-ethnic situations; and the perennial paradigm of the interpreter's invisibility – all these are recognizable within our current discussions on war and conflict.

Nevertheless, this visual context, the experience of identifying the interpreter in a photograph, has also raised some additional questions, questions which relate to the professional relationship of interpreters with their employers, to the spatial positioning of the interpreter, and to the symbolic status that interpreting often assumes in political strategies. The identity of interpreters in war is visually mediated by their clothing, which provides key evidence of the extent of their identification with or absorption into their employers' agendas. Thus Kola and Trini, Smiley's interpreters in Albania, wear clothes indicative of their own ethnic origins rather than of their association with an underground army trained by the British. The military interpreter in the Azores is dressed in the uniform of the RAF, totally integrated into the British war effort, whilst Stalin's interpreter wears something that resembles a formal dress uniform, integrating him into the higher echelons of the service as a worthy interlocutor of world leaders. The interpreter in Figure 10.3 stands in an intermediate position, wearing the uniform of a British regiment but one that, in its subtle modifications, suggests his position as a subordinate insider, not fully accepted within the traditional British army structure.

Photographic evidence also enables us to see something of the relationship of interpreters to the specific geographical and spatial contexts in which they work. Smiley's interpreters are evidently at home in the forests of their own country, whereas the RAF interpreter in the Azores operates from a "home away from home", established in a foreign base with the agreement of the local population, and the interpreter in Figure 10.3 works in a formerly hostile country now occupied by the Allies. Pavlov, as an international interpreter, is positioned in a delocalized hybrid space, a conference venue, deliberately constructed to permit people from different nationalities to engage publicly with each other.

Finally, the evidence of photographs can suggest the ways in which interpreters begin to assume a symbolic status in the visual perceptions of others, both at the time and later. The RAF and Gurkha interpreters, in their different ways, stand as symbols of the new authorities which would come to operate in Europe and Asia in the post-war period. In the case of the most public and politically crucial of the images, the Stalin/Churchill encounter, the interpreter in his partial visibility becomes symbolic of the political strategies of the time, standing proxy for a desired warmth among international leaders – an entente which would barely survive the very meeting itself. Smiley's Albanian colleagues gaze out at us today as typical victim interpreters, paying the price for their association with foreign interveners much as we know men and women have recently done in Iraq and Afghanistan.

As Azoulay argues, these visual images collapse time and importantly bear "the potential for permanent renewal" (2012, p. 27), a renewal often framed by the very ambiguity of their contemporary presentation to us in archives like those of the IWM. The pencil sketch at Figure 10.5, for example, comes to us today through a palimpsest of captions.

The artist, Clifford Charles Brady, serving as an officer with the Chindits in Burma, has identified his drawing as "'Mr Prembahadur' of Kohima, May 23/44". The image is catalogued under "interpreter", with the IWM description: "Mr Prembahadur was an interpreter with a Chindit Column and was awarded an MBE for his service". Does this archival caption mean that a major British distinction (MBE, Member of the British Empire) was awarded specifically for services as an interpreter, and at the time and place noted by the artist? Searching for Mr Prembahadur in the National Archives, however, leaves some uncertainty about this interpreter's identity. In August 1944, a Naik (corporal) called Prembahadur was

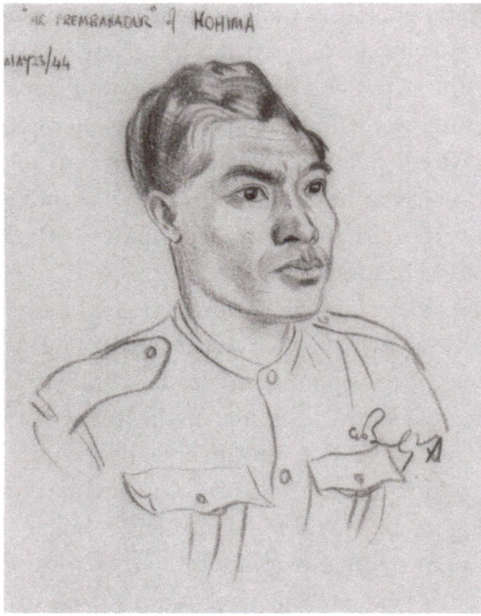

FIGURE 10.5 "Mr Prembahadur" of Kohima, MBE, May 1944

indeed awarded a distinction, but this was the Burma Gallantry Medal rather than an MBE. It is not until 1951 that records reveal a Lieutenant Prembahadur receiving an MBE, with a citation specifically referring to interpreting-related work, although this reference places his interpreting as having been in Malaya between 1948 and 1950,[1] rather than in Burma in 1944. Framed by the uncertain later additions of archivists, Mr Prembahadur, the mystery interpreter, calls out to us from across what Azoulay terms "the civic space" (2008, p. 18) formed by the gaze and by our interrelations within it.

Note

1 National Archives, Kew, WO 373/36, Fort Hertz Area, 10 October 1944; WO 373/129, Citation Lt. Prembahadur, 26 January 1951. "He has gone to great pains to learn the Malay language which he now speaks fluently. On a recent operation in the BAHAU area, without any assistance from an interpreter he worked a surrendered Chinese bandit, extracting from him by degrees the whole enemy layout in the area".

References

Azoulay, A., 2008. *The Civil Contract of Photography*. Translated by R. Mazali and R. Danieli. New York: Zone Books.

——, 2012. *Civil Imagination: A Political Ontology of Photography*. Translated by L. Bethlehem. London: Verso.

Dower, J. W., 2000. *Embracing Defeat: Japan in the Wake of World War II*. London: Penguin.

Farwell, B., 1985. *The Gurkhas*. Harmondsworth: Penguin Books.

Footitt, H. and Kelly, M. eds., 2012. *Languages at War: Policies and Practices of Language Contacts in Conflict*. Basingstoke: Palgrave Macmillan.

Lewinski, J., 1986. *The Camera at War: War Photography from 1848 to the Present Day*. London: Octopus Books.
Porter, P., 2009. *Military Orientalism: Eastern War Through Western Eyes*. London: Hurst and Company.
Pratt, M. L., 2008. *Imperial Eyes: Travel Writing and Transculturation*. London: Routledge.
Rafael, V., 2009. "Translation, American English, and the national insecurities of Empire." *Social Text 101*, 27(4), pp. 1–23.
Smiley, D., 1984. *Albanian Assignment*. London: Chatto and Windus, Hogarth Press.

11

INTERPRETING FOR GENERALS

Military interpreters in Finnish propaganda photography

Pekka Kujamäki

In April 2013, the Finnish Defence Forces opened online access to an archive of around 170,000 World War II-era photographs. As implied by its title, "From the front line to the home front 1939–45" (SA, 2013), the collection covers the war years from 1939 to 1945, encompassing all areas of life during the three military conflicts known as the Winter War (1939–40), the Continuation War (1941–44) and the Lapland War (1944–45). The pictures, taken by photographers of the propaganda companies subordinate to the General Headquarters, appear with captions taken from handwritten reports that were produced after each mission. The reports, comparable to fragmentary notes in travel journals, provide information on the photographer(s) and the location of the events, and short informative captions for individual pictures or an entire sequence. Taken as a whole, the extensive collection constitutes a captivating visual history of a country in military conflict.

For a historian of translation and interpreting, the collection offers around 50 photographs for analysis (based on a search with the word root *tulk** "interpreter", excluding pictures referring to musical interpretation). Even the small sample from this archive shown in Figures 11.1 and 11.2 hints at an interesting visual representation of military interpreting by different actors (soldiers, officers, prisoners of war, women and children) with different ethnic backgrounds (Finns, Germans, Ingrians, Ukrainians) and in different settings (for example interrogation of prisoners at the front or, as in Figure 11.1, in POW camps, and at the early stages of ceasefire negotiations).

The images from front-line interpreting situations illustrate short-term translational spaces (Cronin, 2006, p. 68; Simon, 2012, p. 7) such as radio surveillance or loudspeaker announcements prior to an attack, captured enemy soldiers sitting in groups with an interpreter, or individual POWs being interrogated immediately after their capture. These spaces are set up ad-hoc and are, by virtue of their

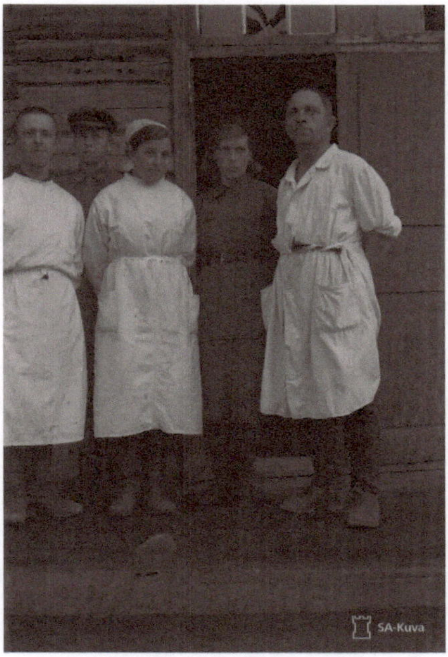

FIGURE 11.1 The Russian personnel of the POW camp hospital. Behind them a young Ingrian who acts as an interpreter. Aunus, 17 September 1941[1]

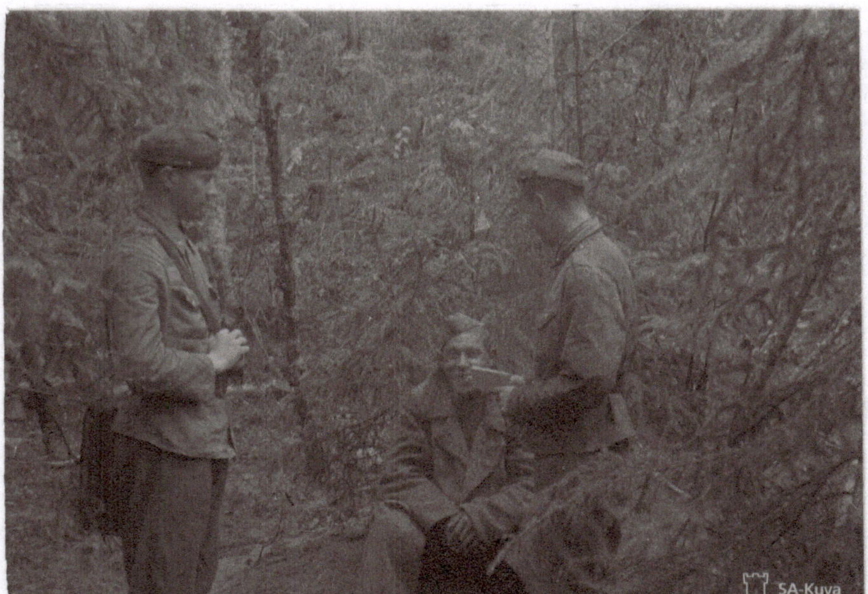

FIGURE 11.2 Major Karttunen, Commander II/45th Inf. Regt, interrogates a defector with the help of an interpreter. Syskyjärvi, 25 July 1944[2]

military context, always framed by prior or anticipated violence. Consequently, through these photographs the viewer is faced with the dust, dirt and sweat of violent combat and with hunger, exhaustion, insecurity and anxiety about the future, but occasionally also, as Figure 11.2 exemplifies, expressions of relief and joy at having survived.

In the published collection as a whole, however, the representation of multilingual military contacts between nations or ethnicities and the mediation practices that evolved around them goes well beyond the 50 photographs that present themselves for analysis through photographers' (often unsystematic) captions. The search for framed military interpreters cannot be based on textual information alone, but must also look at "assumed interpreters" or "assumed translational spaces". By analogy to Gideon Toury's concept of "assumed translation", a target-language text for which there are reasons to tentatively posit a derived, translational quality (Toury, 1995, p. 35) and Lieven D'hulst's concept of "assumed transfer" as a feature of the "interrelatedness of cultural exchange processes" (D'hulst, 2012, pp. 141–42), my concept of "assumed translational spaces" involves identifying the thematic categories and situations in which a framing of multiculturality, multilingual contacts and – consequently – the interpreter's mediating role can be assumed.

This article[3] concentrates on one of the key military events in the Continuation War, an incident that was not only significant enough to be photographed, but also so multicultural and multilingual in character that the presence of an interpreter can safely be assumed: the capture and interrogation of Red Army General Vladimir Kirpichnikov in August–September 1941. After a short discussion of the historical context, I will describe the meeting between the Finnish General and the Russian captive as narrated through the sequence of 19 photographs taken during the event, and examine the physical space and transactional role taken by a third person, assumedly an interpreter. The framing and visibility of the interpreter is illustrated utilizing the "grammar of visual design" of Gunther Kress and Theo van Leeuwen (2006) and is contrasted with the invisibility of the interpreter in the textual accounts of this situation. The visualization of the invisible is supplemented with background information on the interpreter, which shows how wars move people around and give linguistically skilled individuals a kind of agency that they do not have or take outside the military.

Historical background

General Kirpichnikov was the highest-ranking Red Army Officer to be taken prisoner by the Finns in the Continuation War. In the days following his capture on 31 August 1941, he was questioned, first in the 24th Infantry Regiment, then in the headquarters of the IV Army Corps, and was then moved to the Army headquarters in Mikkeli for in-depth interrogation to extract information about the Soviet troops facing the Finnish Army. In addition to the information he provided, Kirpichnikov gave the Finnish high command an important propaganda tool, and the meeting between the Red Army general and the Finnish General Lennart

Oesch, commander of the IV Army Corps, was documented by Finnish Army reporters and photographers. Reports of the successful Finnish manoeuvres on the Karelian Isthmus and pictures of the high-status Soviet captive were soon circulating in the Finnish newspapers, and Kirpichnikov also appeared in a propaganda news flash produced by the Finnish Defence Forces in autumn 1941. The picture of the Soviet general lighting the cigarette of the Finnish general (Figure 11.3) became iconic for Finland's success in the first months of the Continuation War. This picture is still used in articles and books on military history and in Wikipedia, and it is interesting to observe that the captions usually only refer to the two generals enjoying their cigarettes and their conversation (see, for example, Manelius, 2003, p. 67) – thus repeating the photographer J. Taube's archived caption for the whole picture sequence: "General Oesch and Kirpichnikov in conversation." With regard to unique features of the military event and to the propagandist and journalistic purposes assigned to the materials produced in this situation, the caption seems to cover all relevant information for the military and the general audience.

For a translation historian, however, an additional question arises: In what language did their conversation actually take place? Vladimir Kirpichnikov was born in Russia and received his military education in Russian and Soviet military institutions; little is known of his mastery of languages other than Russian (Hristoforov, 2008, p. 545). Lennart Oesch, in contrast, grew up in a multicultural family in Eastern Finland, where he learned German and French from his father, a Swiss

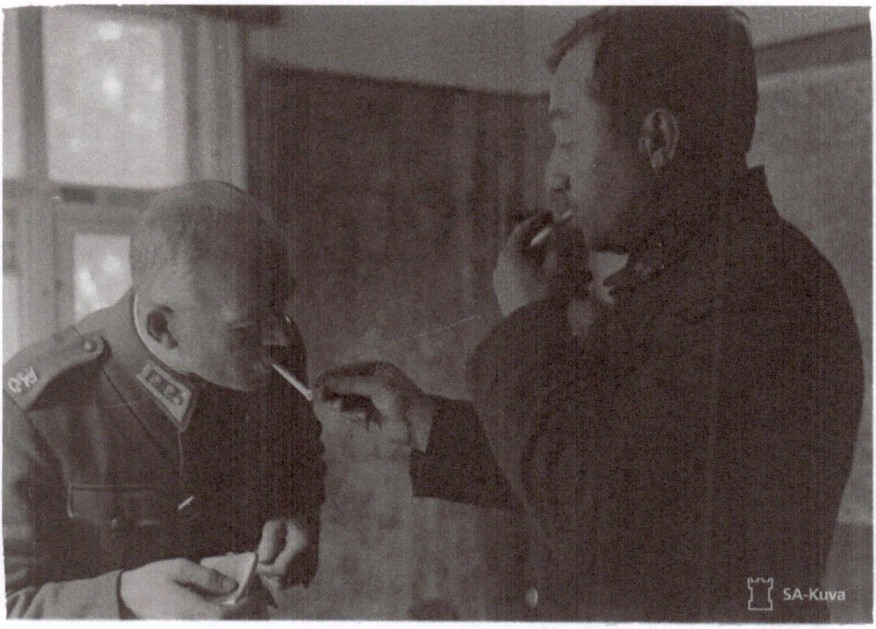

FIGURE 11.3 The captured Red Army General Kirpichnikov lighting Lieutenant General Oesch's cigarette[4]

cheesemaker, and Finnish from his mother and his school environment. He also attended German and French military academies. Oesch knew some English and some Russian, but presumably not enough to conduct the interrogations alone (Seppälä, 1998, p. 85).

A closer look at the 19 pictures of this meeting between the two soldiers supports the assumption that Oesch's linguistic knowledge was insufficient to converse with his captive. It is interesting to see how much physical space an unknown "third person" has been given in the pictures – a person who seems to be in constant proximity to the main characters of the visual narrative (see Figure 11.4). In other words, despite his textual invisibility and unknown identity in the archived captions, in the pictures the figure of the interpreter and his role as a facilitator feature very visibly.

Represented interpreting

Kress and van Leeuwen (2006) analyse images as social semiotic signs that involve two kinds of participants. "Represented participants" are people, places and things depicted in images, while the producers and viewers of images are "interactive participants". In their visual analysis, Kress and van Leeuwen identify three kinds of relations between the two participant groups (p. 114). First, images involve interactions and conceptual relations between the represented participants. Second,

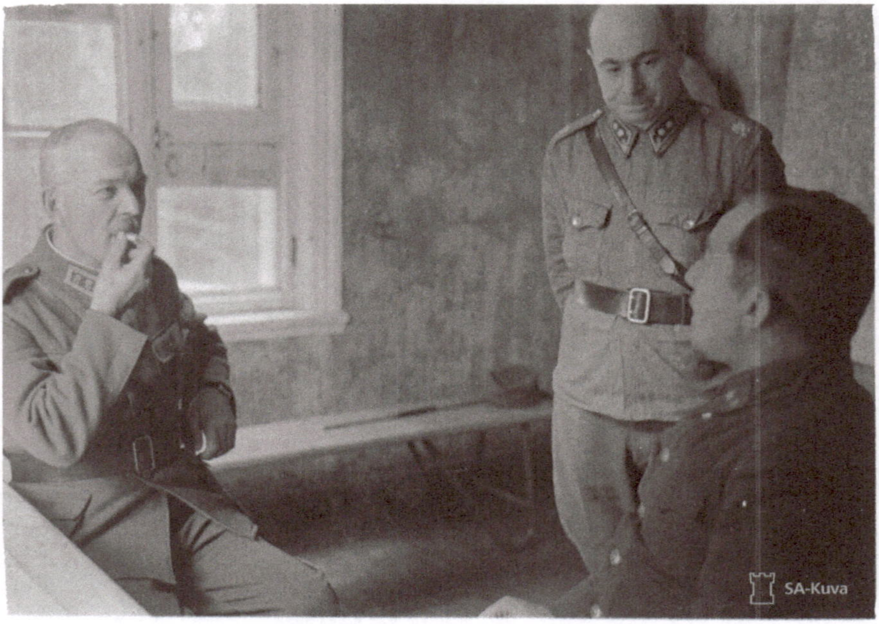

FIGURE 11.4 General Oesch and Kirpichnikov in conversation. Kavantsaari, 2 September 1941[5]

pictures reveal an ideational or imaginary engagement of interactive participants with represented participants. Finally, because images are used as communicative events, they also contain relations between interactive participants, for example a photographer and a viewer. The following analysis concentrates on "representational relations", in other words the relations between the represented human participants, and especially on the represented role of the interpreter in the given historical situation. To analyse these relations, I focus on eyelines and other vectors expressing transaction, that is, the narrative processes between "Actors" (the active participants) and "Goals" (the passive participants) (see ibid., pp. 62–75).

Technically speaking, the sequence of 19 photos is a set of black-and-white pictures taken from an oblique horizontal and an eye-level vertical angle using a medium close up or medium shot, which indicates a photographer following the discussions nearby, as part of the situation but nevertheless not involved. The pictures are mostly rather flat, with the exception of the famous "cigarette picture" (Figure 11.3). There, Kirpichnikov's image is partially surrounded by dark shadows, whereas Oesch's figure stands out in relief against the brighter light coming from the window behind him – thus adding dramatic intensity to the picture. Apart from the first three frames, in which Kirpichnikov is being escorted to the discussions, all the pictures have been taken indoors in a sparsely furnished room with a table, chair and some maps.

In nine of the 19 pictures, the interpreter is featured explicitly (his face or parts of his body can be recognized in the picture) and in three pictures implicitly (through other visual clues to his presence). Taking into account the interpreter's role, the set of pictures could be provided with the following narration:

> The Russian officer has been taken by car to the IV Army Corps headquarters and escorted by intelligence and staff officers, including the interpreter, into the building, where General Oesch is waiting for him with a pack of cigarettes in his hand. The cigarettes are lit, and the two generals sit down to start the conversation at Oesch's desk, the interpreter standing prototypically between them (Figure 11.4). The interpreter, a lieutenant in rank, is wearing the uniform of a Finnish officer. After a short while Kirpichnikov, Oesch and the interpreter continue the conversation in front of a large wall map. The sequence closes with a frame in which the Russian captive is sitting at the table again, while General Oesch and other officers, including the interpreter, stand very closely around him.

A closer look at the composition and at the vectors created by the represented participants' eyelines reveals a set of visual narrative processes that enables more detailed description of the interpreter's role in the discussions. In Figure 11.4, Oesch and the interpreter are the most salient participants: Oesch is seen sitting at an oblique angle, the interpreter as standing at an almost frontal angle, both partly facing the viewer, whereas their counterpart is only partially visible to the viewer and is also slightly out of focus. Oesch's and the interpreter's shared Actor status, as

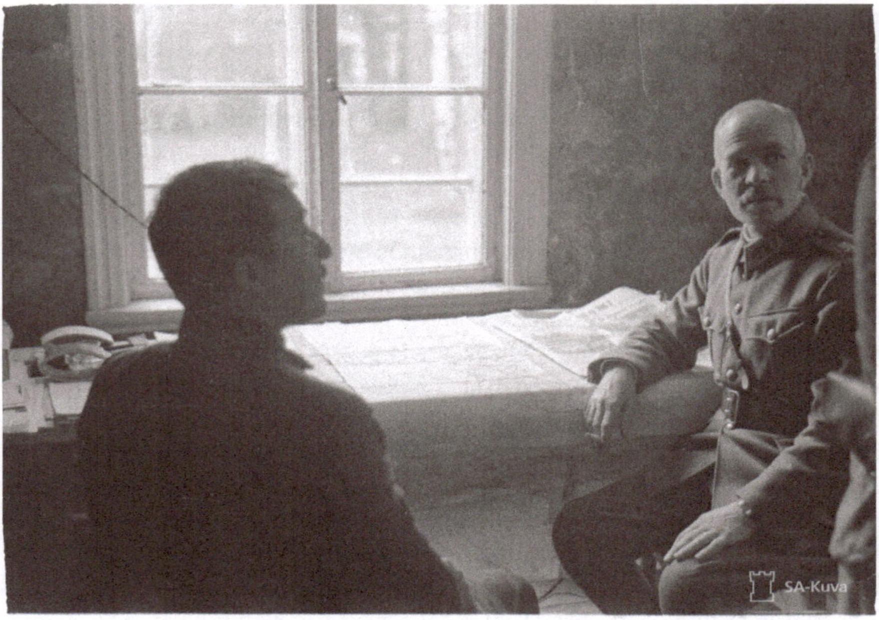

FIGURE 11.5 General Oesch and Kirpichnikov in conversation. Kavantsaari, 2 September 1941

indicated by the composition as well as by their similar uniforms, is strengthened in the way both look at Kirpichnikov, the Goal. This reflects the interrogative nature of the situation, where there is a clear demarcation between "us" and "them", between the powerful and the powerless. In this picture, the interpreter has been represented on the side of the powerful, not as a neutral mediator.

The contradiction between the archived captions to the 19 photographs ("General Oesch and Kirpichnikov in conversation") and the information conveyed by the series is best illustrated by pictures in which the interpreter has not been framed at all.

Figure 11.5 shows two participants sitting at the table, General Oesch facing the photographer and the viewer on the right, General Kirpichnikov with his back to the camera. The vectors running from both participants' eyes point at the same Goal, outside the frame, and reveal the two generals' narrative dependency on a third participant. The invisible subject of the interpreter becomes visible through the transactional relations implied by the generals' gazes.

In Figure 11.6, Oesch and Kirpichnikov are standing in front of the wall map, with the interpreter at Oesch's side partly visible behind him. The eyeline vector from Kirpichnikov to the interpreter and back creates a bi-directional transactional relation between them. Kirpichnikov and the interpreter are thus the represented Actors in this event, and the interpreter has become the actual interrogator of the Russian captive. General Oesch's activity, in contrast, seems to be of secondary importance. The glance at the floor, away from the interactors, as Oesch spits out pieces of tobacco, makes him an Actor in a minor process embedded in the major

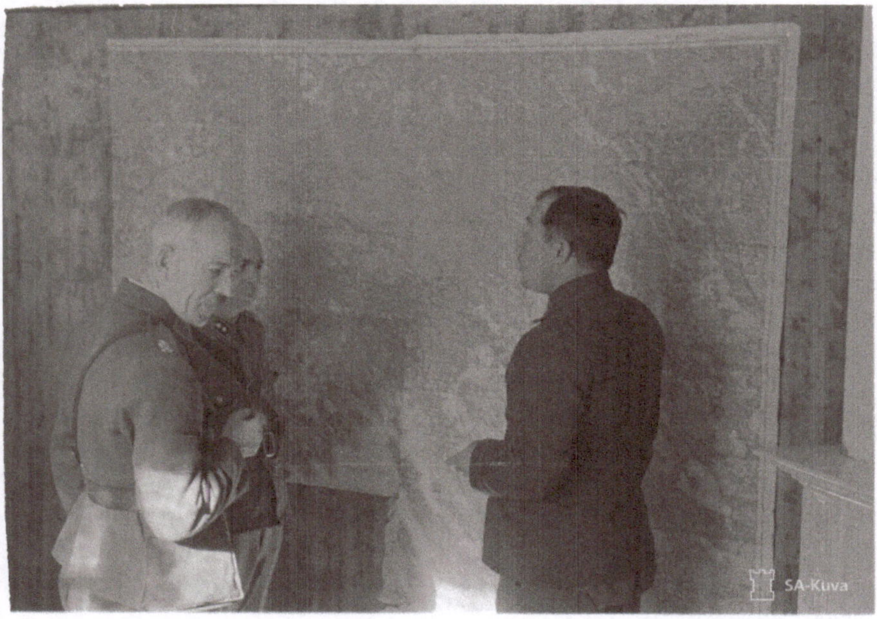

FIGURE 11.6 General Oesch and Kirpichnikov in conversation. Kavantsaari, 2 September 1941

transactional process between the other two represented participants. His role is to wait for Kirpichnikov's information as provided to him by his interpreter.

Identifying the interpreter

While Kirpichnikov, as a uniquely high-ranking captive, appears quite often in recent historical research and journalism on prisoners of war and their use in warfare (see Manelius, 2003; Hristoforov, 2008; Porvali, 2012), his interpreters have attracted much less analytical attention. The existing research highlights the importance of POWs for intelligence and merely implies the significant, yet invisible and anonymous, role of interpreters. In historical research more generally, interpreters are usually dealt with as a faceless and nameless group of people whose language skills make them important for the theatre of war, but who are rarely worth identifying or discussing in any detail. In this sense, the ambivalent discourse around military interpreters clearly reflects the way they are represented and identified, textually and pictorially, in the military archives. Perhaps it is not coincidental that translation and interpreting tend to lie outside the purview of historical research. When a phenomenon is neither physically visible nor mentioned by name in source material, it presents obvious difficulties to the researcher; the phenomenon is most safely mentioned in an aside or as a facet of a larger issue (such as intelligence or the mortality of Red Army POWs), or as an additional capacity or function of a person under scrutiny.

The information we have on Kirpichnikov's interpreters may illustrate this tendency to "sideline". The *habitus* of these "third persons" is represented first and foremost in relation to their activity as intelligence officers; their interpreting practice is noted only in a subsequent sweeping remark. Igor Vahros, for example, is known primarily as an intelligence officer and as one of the developers of Finnish intelligence activities in the Finnish Defence Forces during the Continuation War. He was also the officer most closely involved with General Kirpichnikov, interrogating him, interpreting for him and later translating the memoirs that the Army high command persuaded the general to write for propaganda purposes (Porvali, 2012, p. 108). But before Vahros, as mentioned above, the first intelligence information was obtained from Kirpichnikov in the IV Army Corps by General Oesch and a third person, identified in Igor Vahros's diary entries and letters and in editorial commentaries as Lieutenant Georg Baronin (pp. 109–13). Porvali relates: "Lieutenant General Oesch and Lieutenant Baronin interrogated Kirpichnikov on the locations of Red Army troops and mine fields" (p. 112).[6] Baronin's actual function as the interpreter of these interrogations is only referred to in a short footnote a few pages earlier: "Georg Baronin; typesetter from Helsinki, interpreter of the IV Army Corps and POW interrogator" (p. 109).[7]

Interestingly, Baronin himself defined his *military* habitus or role as that of the military interpreter:

> During the last wars I acted as an interpreter, in the II Army Corps during the Winter War and in the IV Army Corps during the Continuation War. I was involved in interrogations of General Kirpichnikov, who was captured in the pocket of Porlammi during the attacks of the Continuation War. I was promoted to the rank of lieutenant and discharged from service due to my age during the Continuation War.
>
> *(Baronin, 1988, p. 31)*[8]

By the time of Kirpichnikov's interrogations, Georg Baronin was already 46 years old. Born in 1895 in Vyborg, where his father acted as the commander of Castle Vyborg, Baronin went to a Russian school there and later, after his parents' early death, in St Petersburg. After finishing the seven-year classical lyceum, he and his brother volunteered for the army. In 1916 he completed his service in the Russian artillery institute in Hatsina and was promoted to the rank of "deputy praporshchik" (equivalent to an ensign). From May 1916 to October 1917 he fought in the First World War on the Minsk front. After Alexander Kerensky's government was overthrown by the Bolsheviks in the October Revolution of 1917, he resigned from Russian military service because of his Finnish citizenship and volunteer status, and travelled back to Finland. There he enrolled at the Finnish Coastal Artillery to perform his obligatory Finnish military service. After these two tours of duty he moved to Helsinki and worked in a printing house as a typesetter, an occupation that he continued after the Second World War until his retirement in 1966 (ibid., pp. 30–32).

Baronin's own narrative provides a good illustration of the background of many other military interpreters working for the Finnish Defence Forces during the

Second World War. They came from multilingual families in border areas (Eastern Karelia) or multicultural cities (Vyborg, St Petersburg), where Finnish and Russian coexisted as colloquial and official languages. As such, Baronin's background embodies the historical period of Finland's autonomy under the Russian tsar and shows how both this era and the October Revolution contributed to the multilingualism of the Finnish army. Marshal Mannerheim's military education and career in the Imperial Russian Army before becoming the Finnish Commander-in-Chief confirms that Baronin's military background was not unique (see Screen, 1983; Jägerskiöld, 1986). And in the aftermath of the Russian revolutions, Finland received large numbers of Russian emigrants who, along with many other groups from St Petersburg, Eastern Karelia and Ingria (Leskinen, 2012), were later used as interpreters and interrogators – working against their former home country, which had alienated them ideologically.

The autobiographical narrative also shows that Baronin was, like most of his colleagues, a military interpreter in the strictest sense of the term: his interpreting activities started and ended with the wars. There is no mention of earlier experiences or post-war activity as an interpreter. It seems reasonable to assume that Baronin's military background in both armies and ranks, together with his multilingualism, made him an obvious candidate for high-level interpreting tasks in the two army corps. His military background guaranteed terminological knowledge and his bilingualism colloquial credibility – both important assets in sensitive interrogational situations. The specific skills of interpreting were taken for granted. His interpreting probably consisted of the standard tasks of a military interpreter, mainly interpreting during the first field interrogations of captured Red Army soldiers before handing them over to intelligence and – as in case of Kirpichnikov – to the high command in the Army headquarters in Mikkeli.

Conclusion

The capture of General Vladimir Kirpichnikov became a wartime media event in which the propaganda reporters, photographers and film-makers of the Finnish Armed Forces produced a multimodal set of reports, pictures and film inserts of the meeting between the defeated general and his triumphant counterpart. The materials were created by the Finnish military for the home front with the goal of convincing the audience of the successful advance of Finnish troops on the Karelian Isthmus and around Vyborg. Bearing in mind the communicative purpose that initiated the photography in the first place, it seems only natural that the interpreter fell outside the circulated frames and is absent from textual documentation, both in journals and in archival photograph captions. The 19 pictures I have discussed were not taken in order to represent military interpreting and interpreters, yet for a translation and interpreting scholar they can do so nonetheless. The case of the generals' interpreter serves as an example of the special material dimension of interpreters' (in)visibility that needs to be accounted for in further analysis of framed interpreters in archives or collections such as "From the front line to the home front 1939–45" in Finland (SA, 2013): the interpreters are mostly textually invisible, despite their visibility in individual pictures and despite the importance apparently accorded them by certain visual cues related to framing.

In the case of my corpus of photographs, several cues together point to a representation of interpreting that breaks with the common understanding of interpreters as facilitators of a dialogue, standing on neutral territory or in a space "in between" (for a critical discussion of "neutrality" and "in between", see for example Kahane, 2007; Tymoczko, 2010). The sequence of photographs shows the military interpreter Georg Baronin standing between the generals – thus, at first glance, constructing a prototypically neutral space of mediation. However, this neutrality is challenged by the visual narrative processes we can observe in the photographs, which depict the interpreter as an embodiment of the Finnish Defence Forces, an attribution already signalled by his uniform. Additionally, the representation disturbs the presupposition of Baronin's role as an uninvolved mediator in a dialogue between the two generals. Instead, the pictures show Baronin in active engagement as the actual interrogator of Kirpichnikov. This framing also implies a momentary change of power relationships: Baronin has assumed responsibility for the gathering of intelligence, whereas the Finnish general – in textual captions the main actor of the event – has stepped back into the role of a supporting actor.

Notes

1 Here and throughout, all translations from Finnish are my own.
Finnish original caption: "Vankileirin sairaalan venäläinen henkilökunta: Pietarilainen lääkäri, lääkintävääpeli ja kaksi hoitajatarta. Takana inkeriläinen nuorukainen, joka toimii tulkkina."

2 Finnish original caption: "II/JR 45:n komentaja, pataljoonan komentaja majuri Karttunen kuulustelee tulkin avulla erästä yliloikkaria."

3 This article was written under the auspices of the multi-phase project "In Search of Military Translation Cultures: Translation and Interpreting in World War II in Finland", directed by Pekka Kujamäki and funded by the Academy of Finland. For more information, see http://translationinww2infin.wordpress.com.

4 Finnish original caption: "Vangittu punakenraali Kirpitshnikoff sytyttää kenraaliluutnantti Oeschin savukkeen."

5 Finnish original caption: "Kenraalimajuri Oesch ja Kirpitsnikov keskustelevat."

6 Finnish original: "Kenraaliluutnantti Oesch ja luutnantti Georg Baronin kuulustelivat Kirpitshnikovia lähinnä alueen neuvostojoukkojen sekä sulutteiden sijainneista."

7 Finnish original: "Georg Baronin; helsinkiläinen kirjaltaja, IV Armeijakunnan tulkki ja sotavankikuulustelija."

8 Finnish original: "Viime sotien aikana olin tulkkina, Talvisodassa II AKE:ssa ja Jatkosodassa IV AKE:ssa. Olin mukana mm. jatkosodan hyökkäysvaiheessa Porlammin motissa vangiksi jääneen kenrm Kirpitsnikoffin kuulusteluissa. Minut ylennettiin luutnantiksi ja kotiutettiin Jatkosodan aikana iän perusteella."

References

Baronin, G., 1988. "Georg Baronin – RT-2:n perustamisvaiheen upseeri kertoo" ["Georg Baronin – founder of the 2nd Coastal Artillery recounts"]. *Rannikon puolustaja*, 30(2), pp. 30–32.

Cronin, M., 2006. *Translation and Identity*. London: Routledge.

D'hulst, L., 2012. "(Re)locating translation history: from assumed translation to assumed transfer." *Translation Studies*, 5(2), pp. 139–55.

Hristoforov, V., 2008. "Kenraali Kirpitšnikovin sotavankeus ja paluu Neuvostoliittoon" ["General Kirpichnikov's imprisonment and return to Soviet Union"]. In: L. Westerlund, ed. *Sotavangit ja internoidut. Kansallisarkiston artikkelikirja / Prisoners of War and Internees. A Book of Articles by the National Archives*. Helsinki: National Archives. pp. 543–55. www.arkisto.fi/uploads/Palvelut/Julkaisut/SOTAVANGIT%20JA%20INTERNOIDUT_WEB.pdf [accessed 7 April 2014].

Jägerskiöld, S., 1986. *Mannerheim, Marshal of Finland*. London: Hurst.

Kahane, E., 2007. "Interpreters in conflict zones: the limits of neutrality." http://aiic.net/page/2691/interpreters-in-conflict-zones-the-limits-of-neutrality/lang/1 [accessed 7 April 2014].

Kress, G. and van Leeuwen, T., 2006. *Reading Images: The Grammar of Visual Design*. 2nd edition. London: Routledge.

Leskinen, S., 2012. "Questions of interpreters' ethnicity and nationality in wartime Finnish military forces." Paper presented in the workshop *In Search of Military Translation Cultures: Methodological and Conceptual Challenges*, 13 December 2012, Joensuu, Finland.

Manelius, J., 2003. "Mitä sotavangit tiesivät" ["What did prisoners of war know"]. In: M. Karjalainen, ed. *Salaisen sodan sivut: tiedustelua, vakoilua ja salatoimintaa jatkosodassa* [*Pages of the Secret War: Intelligence, Espionage and Secret Operations*]. Helsinki: Maanpuolustuskorkeakoulu, Sotahistorian laitos. pp. 61–72.

Porvali, M., 2012. *Salainen tiedustelija: suomalaisen vakoojaupseerin kirjeet 1940–44* [*Secret Scout: Letters of a Finnish Intelligence Officer 1940–44*]. Jyväskylä: Atena.

SA (Finnish Defence Forces), 2013. *From the Front Line to the Home Front 1939–1945*. Finnish wartime photograph archive. www.sa-kuva.fi [accessed 7 April 2014].

Screen, J. E. O., 1983. *"Våra Landsmän": Finnish Officers in Russian Service, 1809–1917: A Selection of Documents*. Åbo: Åbo Akademi.

Seppälä, H., 1998. *Karl Lennart Oesch. Suomen pelastaja* [*Karl Lennart Oesch: Saviour of Finland*]. Jyväskylä: Gummerus.

Simon, S., 2012. *Cities in Translation: Intersections of Language and Memory*. London: Routledge.

Toury, G., 1995. *Descriptive Translation Studies and Beyond*. Amsterdam: John Benjamins.

Tymoczko, M., 2010. "Ideology and the position of the translator: in what sense is a translator 'in between'?" In: M. Baker, ed. *Critical Readings in Translation Studies*. London: Routledge. pp. 213–28.

12

ORDINARY SNAPSHOTS OF INTERPRETERS AT WAR

A narrative of disaster

Xoán Manuel Garrido-Vilariño

The uses of memory

During the Second World War, science and technology were not employed solely for the advancement of human progress but were at the service of a totalitarian regime whose aim was the mass annihilation of civilians. By analysing three photographs shot by Wehrmacht soldiers, and their handwritten texts, this article aims to contribute to the different ways of understanding the nature of the constrained interpreting activity in World War II. Until recently, German military memory in the form of photographs was managed and kept in albums by the family members of those who regarded themselves as the victors at the time the photos were shot; this is one of the reasons why in the past 70 years we have witnessed only gradual changes in the prevailing political thinking on the politics of memory.

However, today the private photographic heritage is being released through the collecting trade, a process that has also attracted the attention of the research community (see Bopp, 2009). Researchers can take advantage of the current market conditions to access private albums or single elements, making them legitimate historical sources. This allows the memory of the victims to be reconstructed out of photographs taken by, and coming from the memory of, the (assumed) victors. To this end, I will refer to Walter Benjamin's "Theses on the Philosophy of History" (1999b), which emphasizes the specific features of minor historical acts.

Taken from disassembled albums, the photographs under study are separated from their original series and marketed according to the current aesthetic and historiographical rules. This is why only the metanarrative of these records may be known, and any analytical procedure must mainly move within the edges of the almost anonymous photos. Addressing the soldiers' narrative intention involves following the photographer's eye in order to share the fiction he intentionally or unintentionally created. However, in treating these commonplace photographs as historical artefacts, it must be borne in mind that during the approximately 70 years since they were taken, many facts – and many hidden agendas that people of the time did not know or did not want to know – have come to light. Two dimensions thus come to the fore: resistance and collaboration. Both are dimensions of the "real reality" of the interpreter who appears in the photographs. When describing these dimensions, there is no room for suppositions or creating fictitious histories: the formal analysis is based solely on the faces, features and gestures of the participants in the image.

Corpus and historical context

The corpus consists of three photographs, chosen from the ICOTI research collection, in which interpreters appear as a visible element of an interaction. The first photograph (Figure 12.1) shows an interpreter with a white armband posing with a group of German soldiers and a family, somewhere in Poland, according to the handwritten and printed information on the reverse. The second and third photos (Figures 12.2 and 12.3) were taken by the same soldier and show the same group of people, Figure 12.2 being the one in which the interpreter appears more prominently. These two snapshots address the same incident. The amateur photographer sent them home, and probably included them in his war album. No captions are available for these two photographs, but the composition of the snapshot suggests the passive presence of an "assumed" interpreter, in the sense set out by Pekka Kujamäki (in this volume).

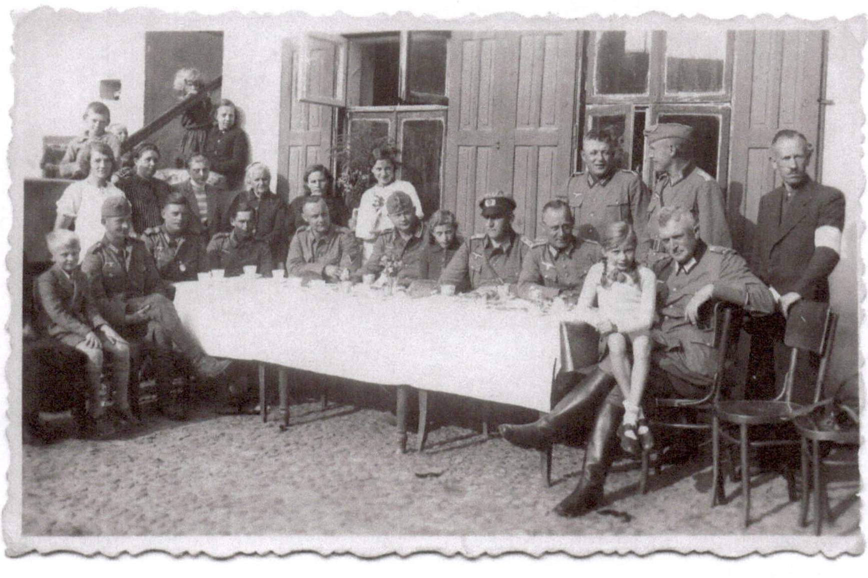

FIGURE 12.1 German family in Poland with members of the Wehrmacht, September 1939

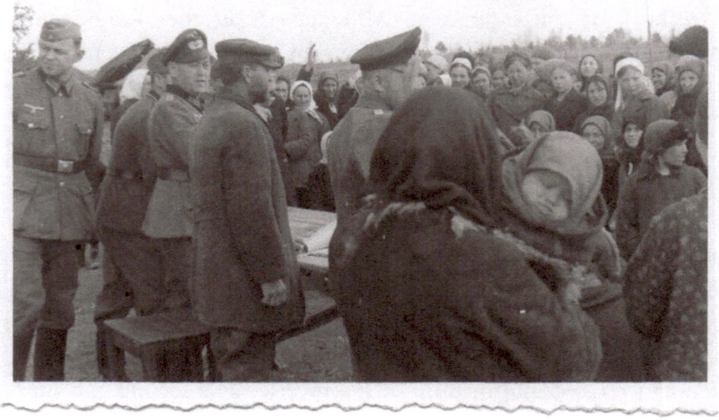

FIGURE 12.2 Eastern Front, Operation Barbarossa, 1941 or 1942

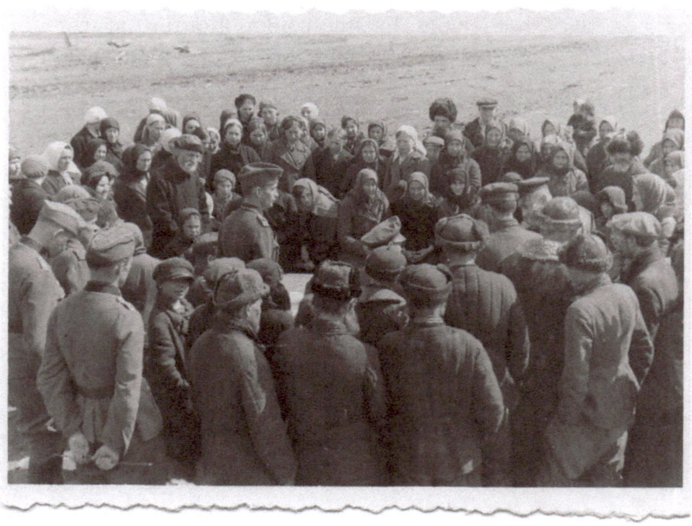

FIGURE 12.3 Eastern Front, Operation Barbarossa, 1941 or 1942

The photo-postcard in Figure 12.1 is dated September 1939, and the only reference to any location is Lodz – from the photography company that developed the image and produced the postcard. On the reverse, the German text reads: "September 1939 'well received' by a German family. The German woman with the coffee pot, the man with the white armband, put themselves immediately at our disposal (interpreter)".[1] This German family in Poland probably helped the

Nazis during their *Polenfeldzug* (military campaign in Poland), which lasted from 1 September to 6 October 1939.

Photographs 12.2 and 12.3 are not dated, but may have been taken after the breaking of the non-aggression pact between Hitler and Stalin. In a first step, I localized the original settings through visual data, namely the landscape, the women's use of kerchiefs around their heads, and the men's clothes and caps. From the clothing, it appears that the photos were shot in Eastern Poland or Western Ukraine, most likely in the region of Galicia. To apply the terminology in use at that time, the participants could either be "Jews" from a *shtetl* or "Slavs". National borders being somewhat flexible in that period, it can be reasonably deduced that the photos were taken in 1941 during the early part of the Eastern invasion, and probably in autumn, still judging by the women's headscarves and the men's hats.

The Nazi Party used propaganda to spread its ideas from the time of its inception until its accession to power in Germany in 1933. Hitler founded the Reich Ministry of Public Enlightenment and Propaganda, led by Joseph Goebbels. The Ministry's mission was to disseminate Nazi doctrine through art, music, drama, books and radio. Propaganda and the media were key elements in the regime's stranglehold on power and, especially, in the justification of its ideological project, which basically consisted of first "Aryanizing" Germany and then expanding German territory to the rest of Europe through invasion. The Nazis' political project implied the conquest and subjugation of "sub-humans" (*Untermenschen*). But even a totalitarian regime needs to justify its projects, both to the domestic public and to the countries it plans to occupy. Thus, Goebbels also opened propaganda offices in the occupied countries. These services offered reconciliation with the population of the occupied countries by portraying a friendly, "correct" German army. Posters were produced affirming Germany's respect for people and property and disseminating imagery of German soldiers as friends of the children.

The year 1933 also saw advances in photographic techniques and the growing popularity of photography as cameras became affordable for the general population. By 1939, some 10 per cent of all Germans owned a camera. As Petra Bopp (2009) notes, describing the photo albums created by the German Wehrmacht soldiers, well-known companies such as Kodak, Agfa, Zeiss Ikon, Voigtländer and Leitz had developed more compact, small-format cameras which were easier to carry and use, and these became the cameras of choice of soldiers in the Second World War. Although taking photos would be prohibited by Himmler after he attended the first mass executions, in 1939 the Ministry of Propaganda was still calling upon the population to continue using their cameras during wartime. In addition to written correspondence, snapshots taken by the soldiers would serve to strengthen the bond between the various fronts and "back home" (Bopp, 2009, p. 40). The snapshots were obviously individually produced, although it should be remembered that they were directed at a collective: their dissemination was probably domestic in scale, and the intention in taking the photo was probably to inform relatives and friends in Germany of the army's accomplishments on the Eastern front, according to the propaganda requirements of the Third Reich.

The "visual tank"

In his 1936 essay "The Work of Art in the Age of Mechanical Reproduction", Walter Benjamin highlighted the photograph's ability to reveal aspects that the human eye does not see (Benjamin, 1999a, p. 214). From this perspective, the camera captures a unique moment of which not even the photographer himself is aware at the time of the shot. It is a moment of truth which would have escaped had it not been revealed by the photograph, but which now becomes a personal memory of the person who took the snapshot and a collective memory thanks to the original's capacity to be technically reproduced. However, by 1939 the camera had become a new weapon of war, as Janina Struk points out:

> The enthusiasm with which photography had been appropriated for the National Socialist cause before the war continued after 1939. When the German army entered Poland they carried not only their guns but also their cameras, important weapons in war. In Eastern Europe photographs were taken, officially and unofficially, of the humiliation of the people and their culture, the Slavs as well as the Jews.
>
> *(Struk, 2004, p. 57)*

Bopp (2009, p. 41) also discusses the wartime uses that the German photographic industry reserved for the camera. She reproduces a leaflet entitled "The visual tank!" (*Der optische Panzer!*), showing how the transition from individual memory to collective memory involved a number of totalitarian ideological elements, with photography used as a tool to exercise domination and violence.

In fact, Benjamin already showed that Fascism tried to use technological advances to organize the proletarian masses, because

> Fascism sees its salvation in giving these masses not their right, but instead a chance to express themselves. The masses have a right to change property relations; Fascism seeks to give them an expression while preserving property. The logical result of Fascism is the introduction of aesthetics into political life. The violation of the masses, whom Fascism, with its *Führer* cult, forces to their knees, has its counterpart in the violation of an apparatus which is pressed into the production of ritual values. All efforts to render politics aesthetic culminate in one thing: war.
>
> *(Benjamin, 1999a, p. 234)*

Benjamin thus warns of the dangers of technical progress, in this case advances in photography, under the control of a totalitarian power. On the one hand, photography allows these powers to control large swathes of the population, while on the other it creates an illusion of freedom and independence in the individual – as with the soldiers in these photo-postcards, who can use a camera to capture a historic moment. Bopp emphasizes the figure of the *Knipser*, an informal German name for the amateur photography enthusiasts whom the Nazi regime trained in mass photography courses with the aim of indoctrinating them in "the unlimited national task of spiritual and mental reconstruction". The regime's invitations in specialist photography magazines featured calls such as: "Put your camera at the service of the people, become a member of the Reich amateur photography club" (Starl, 1995, p. 103; quoted in Bopp, 2009, p. 47).[2]

Photographs taken by private citizens have rarely been considered by historians of World War II, yet the snapshots analysed here prove the historic documentary value of private photographs, which reveal details of the interaction of German soldiers with the native populations, and in the present case how interpreters were deployed to manage communication. Since these photos cannot be analysed as single framed moments, they have to be inserted into the flux of history, as a testimony to the interpreter's work, as a commitment to memory. Memory here is understood in the sense of the duty to remember both the victims – and a significant number of the civilian interpreters in the occupied countries were victims – and indeed all the major and minor acts of mankind, since "nothing that has ever happened should be regarded as lost to history" (Benjamin, 1999b, p. 246).

When Hannah Arendt first published the English translations of some of Benjamin's most important texts, in 1968, she wrote of his historical work:

> [S]trongly influenced by surrealism, it was the "attempt to capture the portrait of history in the most insignificant representations of reality, its scraps, as it were" (*Briefe* II, 685). Benjamin had a passion for small, even minute things.
>
> *(Arendt, 1999, p. 17)*

This is the process I follow in the analysis of the three photographs, considering the major events (the macro history) alongside the minor events (the micro history, in this case the history of the photo) and always keeping in mind that

> there is no document of civilization which is not at the same time a document of barbarism. And just as such a document is not free of barbarism, so barbarism taints also the manner in which it was transmitted from one owner to another.
>
> *(Benjamin, 1999b, p. 248)*

Benjamin denounced the ideological manipulation perpetrated by totalitarian regimes and warned of the dangers both of blind faith in technical progress and of complicity with the victors' concept of history,[3] which rescues both perpetrators and victims:

> To be sure, only a redeemed mankind receives the fullness of its past – which is to say, only for a redeemed mankind has its past become citable in all its moments. Each moment it has lived becomes a *citation à l'ordre du jour* – and that day is Judgment Day.
>
> *(Ibid., p. 246)*

This explains why these photos can be said to be tainted by barbarism. Civilian interpreters were in some cases vital accomplices of the victors, though they were frequently forced to act as interpreters and to play the role of intercultural communicators.

The narrative of the snapshots

These amateur pictures try to introduce an idea of the alleged "benefits" of the Eastern occupation for Germany and the German people. The private, amateur images convey the same narrative as the public images the Nazis used in their propaganda – the private, micro account thus transmits the same idea as the general, macro one given by Nazi institutions. Both the macro and micro accounts can be described as fiction, considering that the reality could be very different. Images 12.2 and 12.3 assert a willingness to communicate with the local population, whereas image 12.1 is an image of what was perceived as a liberation: a German family is being freed from Soviet rule in recently occupied Poland.

In Figure 12.1, a family sits together at a table with members of the German Wehrmacht. It looks very much like a family photo on holiday: the table, covered with a white cloth, has been brought outside from the house. On the table there are coffee cups, a symbol of the warm welcome. The interpreter is distinguished from the rest of the family by the band he wears on his left arm. He stands proudly, well dressed and neatly combed, and overtly poses for the photo, with his hands on the backrest of a chair, drawing attention to the armband and looking fixedly at the camera. All the other people are aware of the photo being taken, but they do not pose in the same way. The Wehrmacht officers look very much at ease, and apparently feel aware of being engaged in an important, but not difficult, mission. The girl sitting on the officer's knee in a way reinforces the family atmosphere; in order to achieve this idyllic setting, it would have had to be known that the male relatives had already joined the ranks of the German army.

With reference to Figures 12.2 and 12.3, the moments framed there give rise to two possible interpretations: either the German army is gathering the population of a village in order to conduct a kind of census (see the desk in Figure 12.2) before separating the "Aryan" from the "non-Aryan", or the officers are explaining the new regime's regulations after announcing the end of Soviet rule in the region. The faces of the people show feelings between bewilderment and uncertainty. A raised hand seems to signify that the villagers are still brave enough to put questions, suggesting that the German army has not yet become a ruthless killing machine (Hilberg, 2003a, pp. 298–99). Yet that assumption should be put in perspective, for the *Einsatzgruppen* (death squads) were already operating in this area and civilians must have been aware of the massacres taking place all over Ukraine, such as that in Babi Yar.

The soldier/photographer of Figure 12.2, in which the civilians include only women and children, is looking towards the woman with the baby in her arms, while the German officer points to the assumed interpreter, the only civilian between the two officers behind the desk. In Figure 12.3, belonging to the same series and shot in the same place, young and old men are also in the group, most of them behind the interpreter, who is turning his back to the camera. Here, too, the officer stares at the interpreter. In the previous picture, the photographer already tried to appear involved with the action by placing the mother and her child in the foreground. In Figure 12.3, however, although most of the subjects are oblivious to the camera, one young boy looks towards the photographer, as a candid witness and adding a sense of everyday life to the composition of the picture.

In these photographs there is no obvious physical action. The interpretation work can be deduced from the German officer's fixed gaze at the interpreter.[4] The way the photographer views the interpreter and the choice of angle are important parameters to measure the ideological effect of this photo. Gunther Kress and Theo van Leeuwen's analysis of the horizontal angle describes how the image producer encodes detachment from the participants (oblique angle) or involvement with them (frontal angle) (Kress and van Leeuwen, 1996, p. 143). As the photographer's decision is determined by his own worldview, his choice often has an ideological effect on the viewer. Thus, the photographer's primary decision in terms of photo composition – the angle – is a crucial question. In Figures 12.2 and 12.3, the photographer is detached from the action, whereas in Figure 12.1 he is involved in it. The producer of the series of two photographs has, if probably unconsciously, aligned himself with the military rather than the civilian population and remained excluded from what is represented. Modern viewers, too, may therefore have no choice but to see the photo's participants as they have been depicted: the soldiers are shown as part of the photographer's world, and the civilian population as the "other".

Figure 12.2 gives a frontal view of the officer looking at the interpreter. The interpreter, seemingly engaged in consecutive interpreting, is shown in profile from his head to his knees. According to Kress and van Leeuwen (1996, p. 144), the positioning of the two bodies generates a double meaning:

> "[A]lthough I am not part of your world, I nevertheless make contact with you, from my own, different world"; or "although this person is part of our world, someone like you and me, we nevertheless offer his or her image to you as an object for dispassionate reflection".

In contrast, in Figure 12.1 the German officers and the interpreter are part of "us". Regarding the composition, the photographer of Figure 12.1 is involved in the moment represented, since he is associated with the victors in the conquest of Poland and the "liberation" of the ethnic Germans. On the other hand, the producer of Figures 12.2 and 12.3 keeps a prudent distance from his subjects. This raises the question of whether the soldier holding the camera had any idea of the destiny of these civilians. In Figure 12.2 the photographer has captured a portrait of the assumed interpreter in the act of mediation. His closed fist reveals the tense atmosphere of these moments, while the interpreter's arm in Figure 12.1 is posed with the hand resting on the back of a chair. The opposition of a resting arm to a fist indicates two attitudes, resistance or collaboration, in these interpreters when faced by the collective for whom they are mediating.

Concluding remarks

Anonymous snapshots may well be considered historic documents, when the moments framed in the pictures go far beyond the frame – when they are inserted into the picture's historical narratives. In the case of World War II, the Nazi state's ideological machine transformed simple anonymous photographers into active agents of manipulation. The first photo, though a private one, is an example of pure propaganda, showing a confident, fearless interpreter who believes that he is performing a dignified job and believes in the ideology he is helping to transmit. The image seems to be well composed, with all the pertinent ritualistic details of the military, the "Aryan" family and the angelic daughter sitting on the officer's lap.

What is striking in these pictorial documents is how the interpreters become witnesses to history in wartime. In a society in turmoil and facing disaster, the interpreter's performance is hampered by the lack of relevant rules or regulations. In Figures 12.2 and 12.3, the constrained interpreter appears caught between the two parties: the Nazi military on one side and the community to whom the message is addressed on the other. As is the case in the second picture, the topics discussed through the civilian interpreter, and thus the very role of the mediator, could be matters of life and death. On the level of his interpreting performance as such, the interpreter is a victim of the permanent tension between the adequacy and the acceptability of the translated message. Faced with the ethical dilemma of being in servitude to power or helping the civilian victims, the interpreter has decided to resist, as his closed fist may demonstrate, which is all the more significant in view of the fact that the snapshot was taken at the beginning of the genocide.[5] At a micro-historical level, photography as a field experience made it difficult to put in practice photography as war propaganda. The basic photographic effects of depth of field and flatness relate the perpetrators to the victims in an assumedly balanced manner. Thus, the narratives show two levels of victimization, embodied by interpreters and moving along an adequacy–acceptability axis through the policies imposed by the military forces: one is driven by the desire to become a victor (Figure 12.1), while the other appears as a victim of history. The previous description of the narratives of the snapshots may thus lead to an in-depth reflection on the policies of memory associated with these images, in such a way that the image of the vanquished[6] interpreters is to be found in precisely the private photographic heritage that was created to feed the image of the victors.

Notes

1 "September 1939 / 'gute Aufnahme' bei einer deutschen Familie / die deutsche Frau mit Kaffeekanne, der Mann mit weißer Armbinde, stellte sich sofort uns zur Verfügung

(Dolmetscher)". Here and throughout, all translations are my own unless otherwise attributed.

2 "Stelle Deine Kamera in den Dienst des Volkes! Werde Mitglied des Reichsbundes Deutscher Amateur-Fotografen!"

3 This term is based on Benjamin (1999b, p. 246).

4 The lack of data makes it difficult to locate Figures 12.2 and 12.3. The military unit wears German Wehrmacht uniforms, and the images show mostly officers. The photographs might have been taken in 1941 or 1942 (personal communication, Gerhard P. Groß, Zentrum für Militärgeschichte und Sozialwissenschaften der Bundeswehr, Potsdam, 5 March 2014).

5 One *Sonderkommando* from *Einsatzgruppe C* was operating in Ukraine at this time (Hilberg, 2003b, p. 1314).

6 Based on Benjamin, as note 3.

References

Arendt, H., 1999. "Walter Benjamin: 1892–1940." In: W. Benjamin. *Illuminations*. Translated by H. Zohn. London: Pimlico. pp. 7–58.

Benjamin, W., 1999a. "The work of art in the age of mechanical reproduction / Theses on the philosophy of history." In: W. Benjamin. *Illuminations*. Translated by H. Zohn. London: Pimlico. pp. 211–44.

——, 1999b. "Theses on the philosophy of history." In: W. Benjamin. *Illuminations*. Translated by H. Zohn. London: Pimlico. pp. 245–55.

Bopp, P., 2009. *Fremde im Visier: Fotoalben aus dem Zweiten Weltkrieg* [*Strangers in Sight: Photo Albums from the Second World War*]. Bielefeld: Kerber.

Hilberg, R., 2003a. *The Destruction of the European Jews*. Vol. I. 3rd edition. New Haven, CT: Yale University Press.

——, 2003b. *The Destruction of the European Jews*. Vol. III. 3rd edition. New Haven, CT: Yale University Press.

Kress, G. and van Leeuwen, T., 1996. *Reading Images: The Grammar of Visual Design*. London: Routledge.

Struk, J., 2004. *Photographing the Holocaust: Interpretations of the Evidence*. London: Tauris.

13

THE VISIBILITY OF COLLABORATORS

Snapshots of wartime and post-war interpreters

Kayoko Takeda

This article discusses interpreters as collaborators and the implications of their visibility, through an examination of two photographs that appear to feature local interpreters in wartime and post-war occupations. The photos are part of the ICOTI research collection, which maintains the captions provided by their resellers. Based on the internal elements and accompanying textual information, the subjects in the photos are probably a Chinese interpreter in a Japanese-occupied region in China during the Second Sino–Japanese War (1937–45) and a Japanese interpreter working in an office of the US occupation forces in post-war Japan (1945–52). In more general terms, they were local interpreters serving foreign military occupiers.

Local interpreters as collaborators of occupiers

In wartime or post-war occupation when the occupied and the occupier speak different languages, a need arises for interpreters who can bridge the communication gap – or, more precisely, who can deliver the occupier's messages and orders to the occupied. Such interpreters may be recruited from within the occupying power or among locals of the occupied land. Due to concerns over loyalty, occupiers may prefer to use their own people, trained in the language of the occupied, as interpreters. But the supply of such interpreters may be limited. In addition, effective occupations seem to rely on the deep knowledge of the language, customs and geography that local interpreters can most readily provide. Local interpreters may also prove useful to occupiers as exemplars of desired behaviour to coax more cooperation from other locals.

In his work on Chinese locals who collaborated with the Japanese army during the war, Timothy Brook (2005, p. 12) states that "occupation creates collaboration".[1] He also cites a similar argument made by Michael Marrus and Robert

Paxton (1995, p. xvi; quoted in Brook 2005, p. 12), in their study of Vichy collaboration with Nazi Germany, that no occupiers "can administer territory by force alone. The most brutal and determined conqueror needs local guides and informants. Successful occupations depend heavily upon accomplices drawn from the disaffected, sympathetic, or ambitious elements within the conquered people". Local interpreters can be considered constituents of such collaboration.

The visibility of interpreters

The visibility of interpreters has been addressed in interpreting studies along two main lines. The first focuses on interpreters' "marked actions" as evidence against the purported invisibility of interpreters in the traditional conduit model. In this framework, the "seeing" party are the researchers, who pronounce interpreters to be visible when they go beyond contributing mere renditions of source speech into a target language. Attention is thus directed to interpreters' interventions, or marked behaviour, during a communicative event as cultural brokers and co-constructors of discourse (see, for example, Angelelli, 2004; Valero-Garcés, 2007).

The other approach focuses on recognition. Here, the "seeing" is performed by all parties, whether present or not, and visibility is determined by whether these parties note the action or existence of interpreters, either positively or negatively. For instance, Mona Baker (2010) discusses the visibility of military interpreters in present-day conflicts, while Kumiko Torikai (2009, p. 1) points out the invisibility of interpreters in historical records, noting that Japanese newspapers have typically omitted interpreters in photo captions of interpreted events with dignitaries. Some researchers suggest that interpreters are generally unrecognized and the only time they become visible is when there is a communication failure or they are alleged to have made a mistake (for example, Apostolou, 2009, p. 4; Torikai, 2009, p. 3).

Adopting this second framework, the present article addresses visibility in terms of interpreters being recognized. As Andrea Brighenti (2007, p. 324) suggests, recognition is a form of visibility that is conditioned by "aesthetics (relations of perception) and politics (relations of power)", adding that visibility can be both empowering and disempowering, and that anything above or below the threshold of "fair visibility" can lead to "distortions in social representations" or social exclusion (p. 330). With attention to the ambivalent, complex nature of visibility, the following sections explore how different parties in the power constellations of occupation may have recognized interpreters who collaborated with the occupiers in the historical contexts of the photos discussed.

Interpreter and traitor

Different captions, different perspectives

A photograph can be interpreted in multiple ways. Its meaning can change depending on when it is seen, by whom, and in what setting. However, the scope

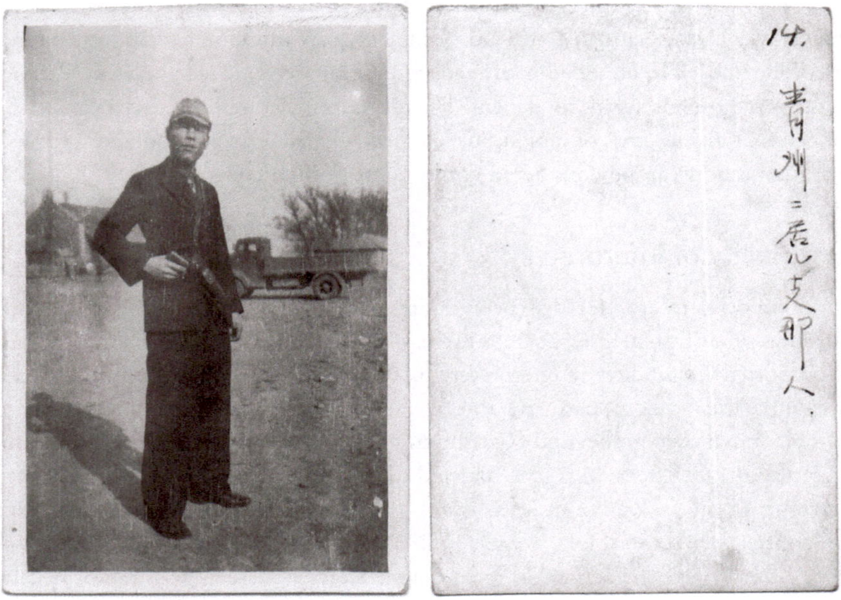

FIGURE 13.1 A Chinese in Qingzhou, 1937–1945

of speculation and imagination can be narrowed to some extent by a caption. As Sunil Manghani (2013, p. 131) suggests, providing a caption to a photo is helpful in “appropriating and re-contextualizing meaning”. Accordingly, one photograph can be explained in different ways if different captions are used. Susan Sontag (1973, p. 109) argues that “even an entirely accurate caption is only one interpretation, necessarily a limiting one, of the photograph to which it is attached”. Figure 13.1 is a case in point, illustrating that a single photo can offer different meanings and perspectives depending on the caption.

This snapshot shows a man in uniform-like clothing and a Japanese army cap. It comes with two sets of textual information. The first is a note written in Japanese on the back of the photo, as shown above, indicating that this is a Chinese man in Qingzhou, China. The other consists of two separate captions in English provided by the reseller of the photo: “Chinese interpreter (Japanese Army)” and “Chinese Traitor”. It is assumed that the reseller had information to warrant such captioning. Each of these captions, while possibly accurate, reflects how its author viewed or wanted to shape the visibility of the subject of the photo.

"A Chinese in Qingzhou"

The note on the back was probably written by the taker and/or original owner of the photo. It reads, “14. 青州ニ居ル支那人 (*Shina* person in Qingzhou). Qingzhou was under Japanese occupation from 1937 to 1945. The word *Shina* originates in the Chinese *Zhina* to refer to China. *Shina* or *Zhina* used to be a neutral term, but it came to carry a derogatory connotation during the Japanese aggression in China (Liu, 2004). Given the numbering, "14.", the note looks like part of the inventory of a photo collection. The photographer was probably a Japanese national in the occupying military and civilian cohort. To maintain its rule over the occupied regions, the Japanese army used locals as go-betweens and puppet administrators (Lary, 2010). "Ruling Chinese by the use of Chinese" was a strategy of the Japanese military at that time (Zhan, 2009). Those Chinese who seemed willing to work with the occupiers were called *tainichi-kyoryokusha* (persons who cooperate with Japan) by the Japanese. Given that he is wearing a Japanese army cap and appears to be posing for a Japanese photographer, the man in the photo may have been a "cooperator" and the photographer may have wanted to document or even commemorate him as such. Gunther Kress and Theo van Leeuwen (2006, pp. 133–40) suggest that taking a photograph from a frontal angle or an oblique angle indicates the photographer/viewer's involvement with the subject or detachment from it respectively. The oblique shot in Figure 13.1 could be understood to mean the photographer's detached relationship with the interpreter, despite his recognition of the subject as a collaborator. The note does not designate the subject as an interpreter, perhaps because his specific role was not particularly remarkable to its author. Despite the pejorative term used in the note, this Chinese man looks rather confident. He does not seem afraid of being recognized in the image.

"Chinese interpreter (Japanese army)"

The first caption provided by the photo reseller identifies the man in the photo as a "Chinese interpreter (Japanese army)". In fact, historical documents indicate that there were a number of Chinese people who worked as interpreters for the Japanese army, especially for the *kenpeitai* (Gestapo-like military police), during the war (Zenkoku Kenyukai Rengokai hensan iinkai [National Federation of Kenyukai editorial board], 1976). If the caption is accurate, this interpreter was probably procured locally – he was one of the occupied who learned the occupier's language and used that knowledge to work for them. During this period, the use of Japanese was mandated for government, military and school functions in the Japanese-occupied regions of China. Japanese was taught at local schools and at language schools established by the Japanese government for specific purposes such as policing. Good command of Japanese became a gateway to a better job and more money (Shi, 2003). These interpreters came from well-educated local elites, and some had also studied at Japanese universities (ibid.; Kawashima, 2007). The interpreter in the photo may have been induced to work for the Japanese army by economic or pragmatic incentives: he may have been given opportunities he could not have enjoyed before the occupation, or maybe he was just trying to stay alive by becoming the indispensable mouth and ears of the occupiers.

"Chinese Traitor"

The other caption provided by the photo reseller reads "Chinese Traitor", offering a strongly negative view of the subject. The English term "Chinese Traitor" may originate in the Chinese word, *hanjian* (a traitor to China). The Chinese concept of *hanjian* specifically refers to a Chinese person who betrays China by collaborating with an external force. During the Second Sino–Japanese War, those Chinese who collaborated with the Japanese occupiers were therefore labelled *hanjian* (Liu, 2000). Chinese interpreters who worked for the Japanese were included among the *hanjian* (Wada, 2009).

All these different captions indicate that the same person, possibly an interpreter, could simultaneously be called *hanjian* (a traitor to China) and *tainichi-kyoryokusha* (a cooperator with Japan), just as La Malinche, Cortés's purported interpreter, adviser and lover, could be seen as a betrayer of the natives, a victim of the conquerors, or a heroine of the new era depending on one's perspective (Romero and Harris, 2005). If we combine the information provided by all the three texts attached to this photo, one way to understand the subject would be that he was a Chinese interpreter hired by the Japanese army and condemned by Chinese locals as a traitor. In post-war trials against *hanjian* and war criminals in China, the prosecutions were based on statements taken from local civilians and former prisoners of war who had been victimized by the Japanese army (Hayashi, 2005; Wada, 2009). Interpreters were among the accused because they were easily recognizable, having been in close contact with victims. Chinese locals may have found it easier to remember the names and faces of Chinese interpreters. It was this visibility, or rather recognizability, that condemned Chinese interpreters as *hanjian* and war criminals. A number of them were found guilty and some were executed.[2] The interpreter in this photograph may have faced the same fate.

"Liberation" and stigma

Interpreters and translators for the occupation forces

Figure 13.2 comes with two separate captions provided by the reseller in English: "Japanese interpreter (US army)" and "11th Airborne, Date: ca 1945, Southern Japan 11th Airborne and Interpreter". Again, it is assumed that the reseller had information on which the captioning was based. Given the specific naming of a US military unit, the photo was probably taken by US Army personnel. It seems to show four Americans in army uniform, four Japanese men in student uniform and one Japanese female interpreter, working together in an office handling transportation arrangements in post-war occupied Japan. Although one of the captions identifies the location as "Southern Japan", all the train timetables on the wall in the background and the historical records of the US Army's 11th Airborne Division indicate that this photo was most likely taken in Morioka or its surrounding area in northern Japan.[3] In terms of Kress and van Leeuwen's discussion (2006) on

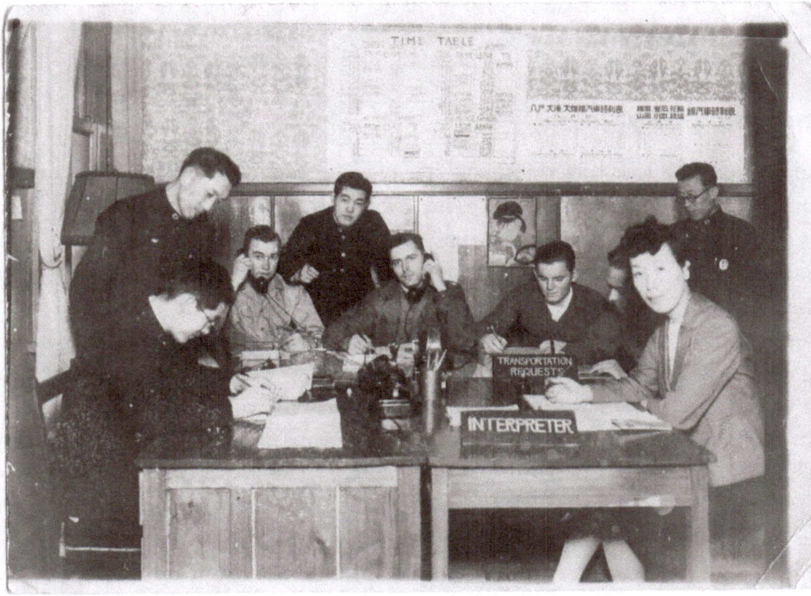

FIGURE 13.2 Japanese interpreter and 11th Airborne

shooting angles, the frontal angle of the US personnel and the oblique angles of the interpreter and other Japanese may represent the photographer's sense of "our world" and "their world" respectively. The interpreter featured in the photo was probably one of the English-speaking Japanese recruited by the occupation powers for various operational roles. Following the landing of the occupation forces at the end of August 1945, finding locals who could translate and interpret between Japanese and English became an urgent and critical task. Without their services, the Supreme Commander for the Allied Powers (SCAP) office could not implement its occupation policies. Although the US Army dispatched thousands of *Nisei* (second-generation Japanese Americans) as language personnel to Japan, post-war language needs were more than they could handle in terms of both quantity and quality (Takeda, 2010). *Asahi Shimbun*, a national newspaper in Japan, carried an article urging male Japanese citizens to apply for interpreting positions with the occupation forces in its issue of 26 August 1945 – 11 days after Japan's surrender and two days before the official arrival of the occupation forces. The Central Liaison Office, an affiliated agency of the Japanese foreign ministry, was established to undertake negotiations and administrative coordination between the occupation forces and the Japanese government. One of its major responsibilities was to procure interpreters and translate documents for SCAP. The Liaison Office's first job advertisement for interpreters appeared in the 1 September 1945 issue of the *Nippon Times*, an English-language newspaper in Japan. Following this urgent ad, the *Nippon Times* started carrying SCAP's "Help Wanted" ads for translators and interpreters almost daily through the first few years of the occupation period. "High pay" was a frequent phrase used in the advertisements to emphasize the pecuniary benefit of working for the occupation forces.

Women in post-war Japan

Initially, most of the advertisements posted by the Japanese and occupation authorities stated "MEN ONLY". The exclusion of women is understandable, given that their military leaders had previously told them to kill themselves when US soldiers landed lest the Americans rape them (Cook and Cook, 2000). This mindset extended to the immediate post-war period, when the Japanese government established facilities for organized prostitution through its Recreation and Amusement Association, allowing "ordinary" Japanese women to be shielded from the advances of occupying troops. There was also a chauvinistic view of women as unworthy to perform "serious" work such as court interpreting in war crimes trials (see Sohma's interview in Torikai, 2009). After a while, however, a dearth of interpreters and translators made SCAP expand its search to Japanese women. Those hired were typically young, educated women, including English-major students. In the midst of devastation, homelessness and widespread unemployment in war-ravaged Japan, these women ventured to work for the former enemy for their own survival and the survival of their families. At the same time, following the oppression and austerity of war, it may have been liberating for some young Japanese women to strike out and earn high pay using their language skills.

Political implications of the composition

Compared to that of the first photograph, the composition of Figure 13.2 appears to be a more deliberate arrangement of each person's position, pose and gaze. It creates the impression that the US military personnel in the middle are in charge, assisted by Japanese nationals in performing the administrative duties of the occupation. The photo may have been taken simply to document US occupation activity in post-war Japan, but it also may have been used for propaganda. Notably, the photo features a Japanese woman fulfilling her professional responsibility and young Japanese men in student uniform working closely with US military personnel. One of them even places his arm on the shoulder of the American in the middle. As the occupation forces requisitioned a university campus as a base for their operations in the Morioka area, students were perhaps easily available as helping hands. However, there may also have been a political aim in making these students and, more importantly, the female interpreter visible through photography: to show a new, democratized Japan with young men and women cooperating with the occupation forces.

The focal position of the female Japanese worker in the foreground, with the sign "INTERPRETER", is particularly significant, as are the captions, which specifically refer to her existence. Her presence and her explicit role as interpreter is made highly visible in this photo, possibly as a representation of Japanese women emerging from traditional, invisible stations to assume new roles. Raising the status of women in Japanese society was considered by SCAP to be an important part of the democratization and demilitarization of Japan. Various initiatives targeted Japanese women in the first few years of the occupation, such as enfranchising women and granting equal rights to women and men in the new constitution.[4] The visibility of the female interpreter should be understood in this sociopolitical context. She was not in a secretarial position, but rather engaged in the vital role of interpreter in what an US Army historian has called "unprecedented cooperation between once bitter enemies" (McNaughton, 2006, p. 415).

Translators and interpreters for the occupation forces performed significant work, such as reviewing and translating evidence for the Tokyo War Crimes Trial or screening and translating Japanese publications and correspondence for censorship by SCAP (Takeda, 2010). However, working for the once hated enemy or collaborating with the occupier seems to have carried a social stigma. Personal accounts of such interpreters and translators, though sparse, indicate that they wanted to keep their experience as "collaborators" to themselves (Akasaka, 2004). They wanted to stay invisible. The interpreter in the photo may have hoped that her image would never travel beyond the US military archives.

Conclusion: The visibility of collaborators

Examining the interpreters in these two photographs illuminates some political dimensions of the visibility of interpreters who collaborate with military occupiers. Acknowledgement as a useful associate by the occupier can be a source of empowerment for interpreters, offering economic and physical security. However, local interpreters are not in a position to choose whether to be visible or invisible. In the asymmetrical power relations of an occupation setting, their visibility is shaped and managed by the occupier. Considered model citizens of the occupied land who speak the language of the occupiers and function as informants as well as messengers to locals, interpreters may be used for propaganda to demonstrate successful cases of local "cooperation", as may have been the case with Figure 13.2. In this context, interpreters' desire to stay invisible is disregarded.

From the viewpoint of the occupied, the local interpreters' visibility carries a different meaning. Being recognized by locals as a proxy of the occupier can be a source of contempt for interpreters, as can be seen in one of the captions for Figure 13.1. Interpreters are not the only collaborators with the occupiers, but they may be particularly easily recognizable as traitors, as they speak directly to locals in the language of evil, oppression and tyranny on behalf of the ruler, occupier and invader. Their "supra-visibility", or visibility beyond the upper threshold of fair visibility (Brighenti, 2007, p. 330), may lead to such extreme cases as interpreters being executed as war criminals or being targets of attack by locals as traitors, as reported in Afghanistan and Iraq (Takeda, 2009).

Local interpreters' motives for assisting the occupation may vary from economic or practical gains to an alleged desire to participate in the occupier's cause.

Collaboration may be motivated by a desire to stay in a position of power, possibly to control and manipulate information to protect locals, or simply to survive. Working for or collaborating with the occupiers is a complex matter, and does not necessarily mean that the interpreters embraced the ideology of occupation. In fact, it is not hard to imagine their struggle with the personal dilemmas and social stigma involved in working against their own country in hostile situations or in assisting the former enemy, who had killed their people and destroyed their land. They may be "just interpreting", but in the context of occupation with its power asymmetries of communication, interpreters can be seen as siding with the ruler and representing them, thus betraying their fellow locals. Conversely, the occupier may not embrace local interpreters as members of their world. The oblique angles from which the interpreters are presented in Figures 13.1 and 13.2 may be an indication of their "stranger" status in the eyes of the photographers/viewers, who are presumed to belong to the occupier's world. These photos may, then, shine a light on the ambiguous, in-between nature of local interpreters' position in the context of foreign occupations.

Notes

1 Brook also discusses Chinese historians' claims that few Chinese collaborated and that their actions were based purely on their personal connection with Japan (2005, p. 3). See Park (2012) for an overview of different views of Chinese collaboration and resistance during the Second Sino–Japanese War.

2 Wada (2009, p. 124) examines three Chinese interpreters tried as *hanjian* and sentenced to six to 15 years in prison by the Shanghai court. Hayashi (2005, pp. 152–53) reports that many of the Korean and Taiwanese war criminals tried in China were interpreters: nine Taiwanese interpreters were sentenced to death or three to ten years in prison (Wada, 2012). The Japanese army had drafted those Korean and Taiwanese interpreters as colonial subjects.

3 Monthly activity reports of the 11th Airborne Division during the occupation of Japan are available in the Modern Japanese Political History Materials Room at the National Diet Library in Tokyo.

4 SCAP even produced a number of movies on women's issues to advance the economic, social and political status of Japanese women (Ikegawa, 2013).

References

Akasaka, M., 2004. "Sengo-naru nichijo, kakusareta shi" ["Post-war everyday life, hidden death"]. *Asahi Shimbun*, 6 June, p. 11.

Angelelli, C. V., 2004. *Medical Interpreting and Cross-cultural Communication*. Cambridge: Cambridge University Press.

Apostolou, F., 2009. "Mediation, manipulation, empowerment: celebrating the complexity of the interpreter's role." *Interpreting*, 11(1), pp. 1–19.

Asahi Shimbun [Asahi Newspaper], 1945. "Tsuyaku to untenshu motomu" ["Interpreters and drivers wanted"]. *Asahi Shimbun*, 26 August, p. 2.

Baker, M., 2010. "Interpreters and translators in the war zone: narrated and narrators." *The Translator*, 16(2), pp. 197–222.

Brighenti, A., 2007. "Visibility: a category for the social sciences." *Current Sociology*, 55(3), pp. 323–42.

Brook, T., 2005. *Collaboration: Japanese Agents and Local Elites in Wartime China*. Cambridge, MA: Harvard University Press.

Cook, H. T. and Cock, T. F., 2000. *Japan at War: An Oral History*. New Haven, CT: Phoenix Press.

Hayashi, H., 2005. *B/C-kyu senpan-saiban* [*Class B/C War Crimes Trials*]. Tokyo: Iwanami Shoten.

Ikegawa, R., 2013. "Senryogun-ga egaita nihon josei-shi: CIE eiga *nobiyuku fujin* no kento" ["A history of Japanese women by GHQ/SCAP: An analysis of the CIE film *Progress of Japanese Women*"]. *Rekishi Hyoron*, 753, pp. 20–34.

Kawashima, M., 2007. "Kahoku ni okeru 'bunka' seisaku to nihon no iso' ["'Cultural' policy in North China and situations of Japan"]. In: K. Hirano, ed. *Nicchu-senso–ki no chugoku ni okeru shakai, bunka henyo* [*Social and Cultural Changes in China During the Sino-Japanese War*]. Tokyo: Toyo Bunko. pp. 61–85.

Kress, G. and van Leeuwen, T., 2006. *Reading Images: The Grammar of Visual Design*. 2nd edition. London: Routledge.

Lary, D., 2010. *The Chinese People at War: Human Suffering and Social Transformation, 1937–1945*. Cambridge: Cambridge University Press.

Liu, J., 2000. *Kankan-saiban: tainichi-kyoryokusha o osotta unmei* [*Hanjian Trials: The Fate of Cooperators with Japan*]. Tokyo: Chuokoron-Shinsha.

Liu, L. H., 2004. *The Clash of Empires: The Invention of China in Modern World Making*. Cambridge, MA: Harvard University Press.

McNaughton, J., 2006. *Nisei Linguists: Japanese Americans in the Military Intelligence Service During World War II*. Washington, DC: US Department of the Army.

Manghani, S., 2013. *Image Studies: Theory and Practice*. London: Routledge.

Marrus, M. and Paxton, R., 1995. *Vichy France and the Jews*. Stanford, CA: Stanford University Press.

Nippon Times, 1945. "Interpreters and guides wanted." *Nippon Times*, 1 September, p. 3.

Park, S., 2012. "Understanding Chinese collaborators during the Sino-Japanese War: review and reflection on the existing literature." *Chugoku Kenkyu Geppo*, 66(11), pp. 27–37.

Romero, R. and Harris, A. N., 2005. *Feminism, Nation and Myth: La Malinche*. Houston, TX: Arte Público Press.

Shi, G., 2003. *Shokuminchi-shihai to Nihongo* [*Colonial Rule and the Japanese Language*]. Tokyo: Sangensha.

Sontag, S., 1973. *On Photography*. New York: Picador.

Takeda, K., 2009. "War and interpreters." *Across Languages and Cultures*, 10(1), pp. 49–62.

——, 2010. *Interpreting the Tokyo War Crimes Tribunal: A Sociopolitical Analysis*. Ottawa: University of Ottawa Press.

Torikai, K., 2009. *Voices of the Invisible Presence: Diplomatic Interpreters in Post-World War II Japan*. Amsterdam: John Benjamins.

Valero-Garcés, C., 2007. "Challenges in multilingual societies: the myth of the invisible interpreter and translator." *Across Languages and Cultures*, 8(1), pp. 81–101.

Wada, H., 2009. "'Kankan-saiban' no jissai" ["Facts about hanjian trials"]. *Chikaki-ni-arite*, 56, pp. 117–26.

——, 2012. "'Sengo Taiwan-jin no 'mibun' ni tsuite" ["Regarding the post-war status of Taiwanese"]. A paper presented at the Seminar on Taiwanese War Criminals, held in Kumamoto, 6 September 2012.

Zenkoku Kenyukai Rengokai hensan iinkai [National Federation of Kenyukai editorial board], 1976. *Nihon kenpei seishi* [*Official History of Japanese Kenpei*]. Tokyo: Author.

Zhan, T., 2009. "Kahoku kairai-seiken" ["Puppet government in North China"]. *Chugoku 21*, 31, pp. 129–46.

PART V

Cold War

The field, the table and the booth

14

INTERPRETERS AT THE EDGES OF THE COLD WAR

Jesús Baigorri-Jalón

Building the history of interpreting requires the use of a great variety of records, including photographic images. This paper focuses on two photographs of events that can illustrate the presence of interpreters at the edges of the Cold War: a luncheon between Marshal Zhukov and Field Marshal Montgomery in November 1945, four months before Churchill's "iron curtain" speech in Fulton, Missouri, and an official meeting between President Kennedy and Chancellor Adenauer in April 1961, four months before the erection of the Berlin Wall by the German Democratic Republic.

This paper applies a historian's perspective to recreate from the present the dynamic role played by the interpreters represented in the two static frames, which appear as frozen scenes. The instant shown by each photographic image is part of a sequence of events (Burke, 2001, p. 143), where the interpreters act as links in the dialogue, while the photographers are filters between the place and moment in which they opened the camera shutter and their potential audiences, including viewers today.

Photographs are non-conventional historical sources and, as Peter Burke (p. 13) says, "uses of images by historians cannot and should not be limited to 'evidence' in the strict sense of the term", because "images allow us to 'imagine' the past more vividly", even if in photography what is left outside the frame could be considered non-existent (Díaz Barrado, 2012, p. 146). As researchers in the history of interpreting, we can also contribute to filling its gaps by making use of past images from a present-day standpoint. Images in general and photos in particular "record acts of eyewitnessing", but they are "mute witnesses and it is difficult to translate their testimony into words" (Burke, 2001, p. 14). Or as Kroes (2007, p. 68) puts it:

> They speak no words, use no rhetorical flourish, no linguistic embellishments or evasions. They freeze transient motion into lasting stillness. … Photographs

> come to us like documents from "the other side", beyond time, beyond life. They are like testaments, last wills drawn up in the service of memory.

The present essay tries to give voice to those images by reflecting on the roles played by the interpreters represented in them, assuming that photos are irreplaceable sources for recreating both individual and collective life memories (Kossoy, 2002, p. 133). Since photographs are subject to manipulation like any other source (Kossoy, 2006, p. 27), historians should be "perennially on guard for the ideology behind the image" (Porter, 1988, p. 200) and apply source criticism as with any other historical records.

Four elements should be taken into account in this analysis, according to Burke (2001, pp. 187–88): the type of representation intended by the image-maker; the contexts in which the image is placed; the fact that series of images give a wider perspective than individual pictures; and the need to read between the lines and use the images "as clues to information which the image-makers did not know they knew, or to assumptions they were not aware of holding".

Events are "historical only insofar as they have a significant and meaningful temporal relationship to earlier and later facts and events" (Rüsen, 2005, p. 130). Thus, these documentary photos could be seen as a mere echo of two news *stories*, but the *history* around and behind them relates to the Cold War, a convenient term based on which the United States and the Soviet Union defined the world's ideological cartography for over four decades.

Distant voices and a still life: Interpreting Marshal Zhukov and Field Marshal Montgomery[1]

An Associated Press (AP) photographer shot this picture (Figure 14.1), limiting the characters to the four people in the photo: Soviet Marshal Zhukov and British Field Marshal Montgomery, flanked by two anonymous junior members of their armies. The one on the right, a British artillery captain identified as the interpreter,[2] is mediating between the military leaders, while Zhukov's interpreter is following a different (unseen) part of the scene.[3]

If we divide the photo with a diagonal line from the upper left to the lower right corner, we see on the right the major object of the photographer's attention – the two main characters with their assistants – and on the left a kind of still life with a fruit bowl, refined glassware and crockery, and two bottles as testimonies to the *Lebenswelt*, which includes "inanimate objects, faces, crowds, people who intermingle, suffer, and hope, … life as we commonly experience it" (Kracauer, 1969, p. 58). The visible part of the label on one bottle reads *Sovietskoe shampanskoe*, hinting that the Soviet Union is hosting the banquet, an assumption confirmed by the caviar on the British interpreter's plate. This luncheon fits well with a description of the Allied Control Council monthly meetings by Montgomery's aide-de-camp,[4] when "Zhukov had the habit of serving a generous buffet lunch with plenty of caviar and vodka" (Henderson, quoted by Axell, 2003, p. 210),

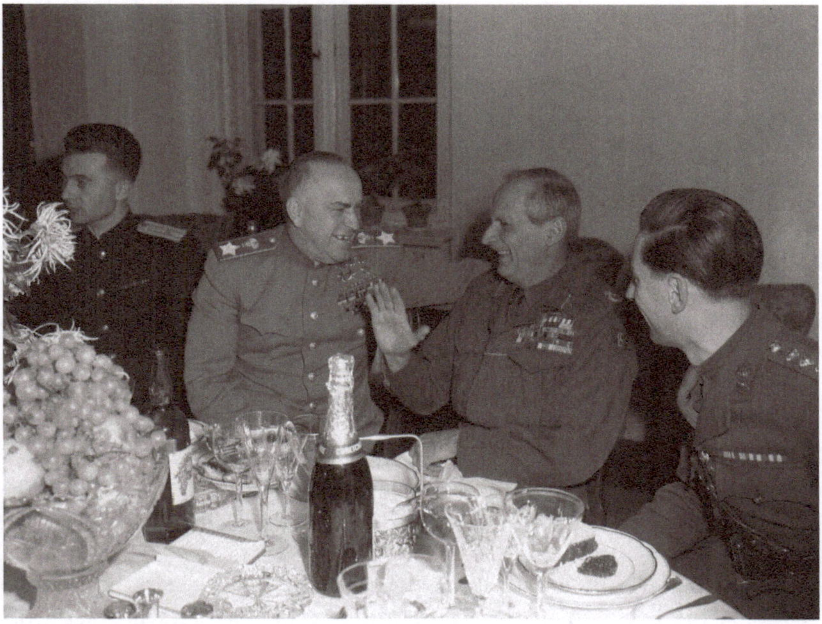

FIGURE 14.1 Field Marshal Sir Bernard Montgomery and Marshal Georgy Zhukov, 18 November 1945

even though Montgomery was famous for his frugal eating and drinking habits (Montgomery, 1958, pp. 359–60, 363).

In terms of Burke's framework, the image sends coded messages that may have exceeded the photographer's intention. Assuming the target audience were readers of the English-speaking press, including the armed forces in Germany, the photo can be seen as a cultural expression of Russian hospitality towards their Allies, but it admits other interpretations. Many viewers may have read the photo between the (military) lines, through the lens of their stereotypes about the other Allies, and perhaps experiencing a syntactic inconsistency between the lavish table and the surrounding landscapes of destruction, hunger or death. It seems clear that their individual or collective judgement about the image will have differed vastly from current readings, made after the passage of time, by less concerned observers, with the benefit of hindsight and in high-definition quality.

The content of the leaders' conversation is not known. In this image, Zhukov and Montgomery are laughing at a remark presumably made by Montgomery. The interpreter's smile reveals the funny nature of the comment as he turns it into Russian, and humour is not the easiest thing to pass between cultures and languages. It is likely that after this lunch, as on other occasions, "[t]here were many speeches and we developed into a sort of mutual congratulation society" (Montgomery, 1958, p. 380), a new potential challenge for the interpreters.

Montgomery and Zhukov are not posing for the photographer. Zhukov's body language – with his left arm stretched over the back of Montgomery's chair – is one of comradeship and bonhomie, something natural in Russian proxemics but less so in British interpersonal relations. We know the two military leaders had been at odds over the interpretation of the Yalta provisions regarding the occupation of liberated territories in East Germany when they were establishing the Control Council in occupied Germany in June 1945 (ibid., pp. 377–79). Even at the time of this photo, disagreement between them was bitter:

> In November, just when I was thinking we were beginning to get our affairs in the British Zone into good working order, Zhukov circulated a memorandum to the Control Council in which he accused me of retaining organised units of the former German Army in the British Zone.
>
> *(Ibid., p. 403)*

The discrepancies expressed by Montgomery may be attributable in part to the differences between the Soviet and Western Allied negotiating styles, which was an ingredient of the Cold War ideological divide and a difficult cultural barrier for interpreters to overcome.

The interpreter who is translating in the photo is the necessary bi-directional link between the two leading characters, who are looking at each other, not at him. The interpreters are sitting at the table with their principals, sharing – time permitting – the same food and drink: a highly symbolic element to frame the interpreters' status. It seems plausible that the British interpreter reached his position, as a trusted "1st Class" high-level interpreter embedded in a liaison or General Staff unit, through selection and accreditation exams in the military.[5] Apart from linguistic skills, each party would require from its interpreters absolute loyalty to their principal. Charles W. Thayer addresses this combination of private conversation between two public figures in the presence of the press (1959, p. 107), where interpreters should possess diplomatic tact and a disposition to act as cultural brokers or, if necessary, as gatekeepers.

Neither Montgomery nor Zhukov mention interpreters in their memoirs. The interpreters' elision or anonymity is commonplace in most written records (Roland, 1982, p. vi). Writing on the translator's invisibility, Lawrence Venuti (1995, p. 2) notes: "The more fluent the translation, the more invisible the translator", while Icíar Alonso-Araguás and Jesús Baigorri-Jalón (2004, p. 130) turn these words around to argue that the more invisible the interpreter, "the greater the illusion of communication without mediation" – as if interpretation were an embarrassing procedure that hinders rather than facilitates smooth communication. Michael Cronin (2006, p. 73) has referred to the concept of the interpreters' inaudibility in mostly symbolic terms, but interpreters were also inaudible in the purely technical sense of the word until voice-recording devices became available in the twentieth century (Baigorri-Jalón, 2006, p. 102). When dealing with images, metaphoric interpretations may be applied, but visibility is understood primarily in real, physical terms. Photographic images become pieces of evidence, or at least traces of the interpreter's role as a critical link in numerous settings. The interpreter framed in this particular photo is not identified, but his mention in the caption invites viewers to think about a function without which the conversation between Zhukov and Montgomery would have been impossible. That presence in the image makes photos all the more significant as major historical sources for interpreters' visibility.

Between stars and stripes: The interpreter's hat[6]

The photo (Figure 14.2) shows President John F. Kennedy meeting with the West German Chancellor Konrad Adenauer in the Oval Office of the White House, Washington, DC, on 12 April 1961 at 9.57 a.m. Between the two men is interpreter Nora Lejins, chief of Language Services at the State Department. At that time, President Kennedy (1917–63) had been in office for only 100 days. Irrespective of the issues discussed in this conversation, the Cold War was a Sword of Damocles hanging over the President's head. That very day, Kennedy told the press categorically that the United States had no intention of intervening in Cuba, but five days later US troops landed at the Bay of Pigs, an event that foreshadowed the October 1962 missile crisis. Four months after this photo was taken, the Berlin Wall, probably the Cold War's most powerful icon, was erected. In November 1963, Kennedy was assassinated and became a legend. Current observers of the photo can add these details by looking at the past and its future from the present, but immediate viewers of this official photograph could only see it as an illustration

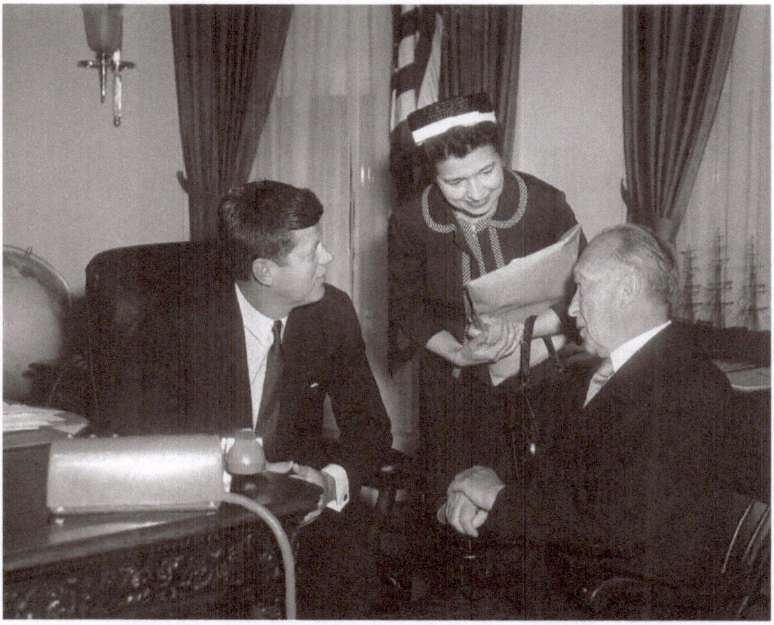

FIGURE 14.2 President John F. Kennedy meets with West Germany's Chancellor Konrad Adenauer, 12 April 1961

of a contemporary diplomatic event. These are the unsuspected possibilities that historical images give us, possibilities of revisiting the past and deepening our views on historical discourse (Díaz Barrado, 2012, p. 161).

The photographer, Abbie Rowe (1905–67),[7] took a shot in which the two main characters are sitting at the corner of President Kennedy's desk, where a sophisticated telephone can be seen – the hotline to Moscow would be established only in August 1963. The composition of this shot is triadic: the interpreter is standing between her principal and his guest, with the curtains and US flag as a backdrop. None of the characters is looking at the photographer. Instead, a sort of triangular gaze flow is established among them, indicating interaction and verbal transcoding through the visual process (Kress and van Leeuwen, 2006, pp. 55, 89). Adenauer appears as an aged, tired-looking gentleman, with his body leaning on the back of the armchair and his hands resting on his walking stick. He is addressing his host, and the interpreter is listening attentively. Kennedy, showing a radiantly photogenic face,[8] is sitting with his body leaning forward, his elbows resting on the arms of the chair and his hands crossed. He is not looking at his guest, but carefully watching his interpreter, the person he understands. The interpreter is standing, slightly bent down – surely to get and provide optimal sound quality. In her right hand she grasps several pencils, which she will presumably use in the meeting as a tool for consecutive interpreting and for the draft memorandum of conversation. She is also holding an envelope, perhaps containing relevant documents on the subject to be discussed, and her handbag is hanging from her left arm. She is wearing a hat with a white band that seems to prolong the stripes of the US flag behind,[9] and she is smiling, so that at first sight she looks more like a waitress taking an order than the State Department's chief interpreter. A visual parallel can be found between the white strip on the telephone – a reflection of the camera's flash – and the band on Mrs Lejin's hat, perhaps suggesting a symbolic link between the oral communication functions attributable to the device and to the interpreter. It seems clear that this was not the arrangement for the whole of the 90-minute meeting, but rather a fleeting instant for the official press photographers (at least two were present).[10]

It is interesting to note here that the interpreter is framed between her principal and his guest, since other photos of the same event and series omit the interpreter. The fact that a woman is playing the role of main interpreter at a summit between two male heads of state or government is a crowning moment in the history of a language service that was established in 1789 by Thomas Jefferson. It anticipates the higher status and representation in the conference interpreting profession that women would acquire in the subsequent decades.

According to the memorandum of conversation on Adenauer's visit to Washington, drafted by Nora Lejins,[11] the subject of the conversation was the balance of payments problem. The participants also included Mr Heinz Weber (Federal German interpreter),[12] who appears between the two statesmen in other photos.[13] We can assume that both interpreters participated actively in the meeting, presumably in both directions: into English and into German.[14]

At the time of the photo, the interpreter Nora Müller Lejins had risen to lead the Language Service at the State Department after a long career in the US Administration. She was born in Cologne in 1913, became a naturalized American (Obst, 2010, 122) and died in 2006 at the age of 93. According to the 1959 *Biographic Register* of the US Department of State, Lejins was at the pinnacle of her interpreting career when the photo was taken. Supervisor of translators for the Navy and translator at the State Department from 1948, she became an interpreter in 1952 and assistant chief of the Language Services in 1957, and was chief interpreter at the time of the photo.[15] Lejins had interpreted for President Eisenhower[16] and would continue to interpret for President Kennedy. She had met her future husband, Peter Lejins, at the University of Chicago in 1934. They married and moved to Latvia "until 'the Soviets took over in 1940, when we got out by the skin of our teeth,' she said, and made their way to the Washington area" (Raver, 1996).

The chief interpreter's biography should be seen in the perspective of a 200-year-old service providing "interpreting and translation services for the US President and Secretary of State, and other senior officials in the White House and Department of State" and "other Departments and federal agencies" (Sawyer, 2008, p. 268). As chief of the Interpreting Division, Nora Lejins performed not only interpreting duties but also a great variety of other tasks, including the selection and training of other interpreters.

Closing remarks

All photos are partial, in the sense that they frame a specific target, which is only a part of the whole, and they represent a past moment – even when developed and shown immediately after the click of the shutter. These two photographs, where military and diplomatic interpreters were included – deliberately or not – by photographers in their version of a reality they witnessed, contribute to the visibility of interpreters and their roles. Potential contemporary viewers probably focused their attention on the well-known characters in the images, but viewers would also have become indirectly aware of a function that was socially unclear, so the particular *framing* of the interpreters helped to reinforce their social visibility. For present-day researchers these photographs, duly catalogued and identified as historical records, are valuable sources. With them, we may fill many gaps in the history of interpreting by searching for historical meaning in silent images that represent scenes of communication. They can be used as a means to reflect on the interpreter as a channel both of the conversations' potential linguistic content and of the extra-linguistic aspects, such as cultural issues, ideological opinions and stereotypes, negotiating styles, body language, spatial scales and distances, or humour – that is, all the contextual and "between-the-lines" elements that observers can imagine, as Burke suggests. The two photos can be seen as visual metaphors of another metaphor, the Cold War ideological construct. Although still at a fledgling stage in 1945, this construct became a chronic element not only of bilateral meetings like the ones in these images, but also in multilateral settings such as the United

Nations, where the main Cold War powers confronted each other with speeches, rather than guns, at negotiation tables and mostly through interpreters.

Notes

1 Caption: "Field Marshal Sir Bernard Montgomery and Marshal Georgy Zhukov, members of the Allied Control Commission, sharing a great joke explained by an interpreter, right, during a luncheon which followed the postponement of a meeting of the Commission in Berlin, on Nov. 18, 1945, owing to the absence of General Dwight Eisenhower and General Marie Pierre Koenig, the other two Allied leaders" (AP Photo).
2 Possibly Hugh Lunghi, www.telegraph.co.uk/news/obituaries/10705902/Hugh-Lunghi-obituary.html [accessed 2 May 2014].
3 In another AP photo, taken a moment earlier, the Soviet assistant's face is turned to Zhukov, with the air of having made a recent comment. This confirms Burke's point (2001, p. 188) that a series of images offers a wider perspective than a single shot.
4 Established in August 1945, this body coordinated the four zones into which Germany was divided, one under each administering Power. It issued recommendations and directives in many fields, but each Power retained a great deal of autonomy in its respective area. The last time a Soviet representative attended the Allied Control Council meetings was in March 1948. By then, relations among the Western Powers and the USSR had deteriorated considerably. Increasing differences culminated in the creation of two separate republics, the German Federal Republic and the German Democratic Republic, a division that was one of the most salient symbols of the Cold War era.
5 For a current perspective on British military linguists, see Lewis (2012).
6 "President John F. Kennedy Meets with West German Chancellor Konrad Adenauer." Abbie Rowe. White House Photographs. John F. Kennedy Presidential Library and Museum, Boston.
7 US Department of State, "Artists: Abbie Rowe and Dick Demarsico", http://art.state.gov/artistdetail.aspx?id=103016 [accessed 7 April 2014].
8 Barthes (1957, p. 180) uses the term *photogénie électorale* with reference to politicians' campaign portraits.
9 "Mrs. Lejins, a German-born linguist, spent most of her working years in charge of language services at the State Department. (There are pictures of her on the wall, at the White House, with John F. Kennedy and German Chancellor Konrad Adenauer and others. *She wore stylish hats.*)" (Raver, 1996; emphasis added).
10 President J. F. Kennedy Library, symbols KN-C17521 and KN-C17522. A similar photo to the one described here was published at the *Yonkers (NY) Herald Statesman*, 13 April 1961.
11 U.S. Department of State: Office of the Historian, Foreign Relations of the United States, 1961–63, vol. IX, Foreign Economic Policy, document 44, http://history.state.gov/historicaldocuments/frus1961–63v09/d44 [accessed 7 April 2014].
12 Ibid.
13 JFK Presidential Library and Museum, "President John F. Kennedy Meets with West German Chancellor Konrad Adenauer", www.jfklibrary.org/Asset-Viewer/Archives/JFKWHP-AR6501-D.aspx [accessed 7 April 2014].
14 "In my entire career, I never encountered a better English-German diplomatic interpreter than Heinz Weber" (Obst, 2010, p. 127). Weber interpreted Kennedy's speech in his visit to Berlin in June 1963, when the President pronounced, with his Boston accent, the famous *Ich bin ein Berliner*, which was (better) understood when the interpreter repeated it in native German (Daum, 2008, pp. 141–42).
15 Department of State of the United States, *The Biographic Register 1959*, p. 424.
16 See, *inter alia*, US Department of State: Office of the Historian, Foreign Relations of the United States, 1958–60, vol. IX, Berlin Crisis 1959–60; Germany; Austria, document 248, http://history.state.gov/historicaldocuments/frus1958–60v09/d248 [accessed 7 April 2014].

References

Alonso-Araguás, I. and Baigorri-Jalón, J., 2004. "Iconography of interpreters in the conquest of the Americas." *TTR*, 17(1), pp. 129–53.

Axell, A., 2003. *Marshal Zhukov. The Man who Beat Hitler*. London: Pearson-Longman.

Baigorri-Jalón, J., 2006. "Perspectives on the history of interpretation." In: G. L. Bastin and P. Bandia, eds. *Charting the Future of Translation History*. Ottawa: University of Ottawa Press. pp. 101–10.

Barthes, R., 1957. *Mythologies*. Paris: Seuil.

Burke, P., 2001. *Eyewitnessing: The Uses of Images as Historical Evidence*. Ithaca, NY: Cornell University Press.

Cronin, M., 2006. *Translation and Identity*. London: Routledge.

Daum, A. W., 2008. *Kennedy in Berlin*. Translated by D. Geyer. Cambridge: Cambridge University Press.

Díaz Barrado, M. P., 2012. "La imagen en el tiempo: El uso de fuentes visuales en Historia" ["Image in time: the use of visual sources in history"]. *HAO* (*Historia Actual online*), 29, pp. 141–62. www.historia-actual.org/Publicaciones/index.php/haol/article/view/774/627 [accessed 7 April 2014].

Kossoy, B., 2002. *Realidades e ficções na trama fotográfica* [*Realities and Fictions in the Photographic Plot*]. Cotia: Ateliê Editorial.

——, 2006. "Imagem fotográfica e história" ["Photographic image and history"]. www2.uol.com.br/historiaviva/artigos/imagem_fotografica_e_historia.html [accessed 7 April 2014].

Kracauer, S., 1969. *History: The Last Things Before the Last*. New York: Oxford University Press.

Kress, G. and van Leeuwen, T., 2006. *Reading Images: The Grammar of Visual Design*. 2nd edition. London: Routledge.

Kroes, R., 2007. *Photographic Memories: Private Pictures, Public Images, and American History*. Hanover, NH: Dartmouth College Press.

Lewis, J., 2012. "Languages at war: a UK Ministry of Defence perspective." In: H. Footitt and M. Kelly, eds. *Languages and the Military: Alliances, Occupation and Peace Building*. Basingstoke: Palgrave Macmillan. pp. 58–69.

Montgomery, B. L., 1958. *The Memoirs of Field-Marshal the Viscount Montgomery of Alamein*. London: Collins.

Obst, H., 2010. *White House Interpreter: The Art of Interpretation*. Bloomington: Author House.

Porter, R., 1988. "'Seeing the past': the English satirical print, 1600–1832 by Michael Duffy; John Miller; J. A. Sharpe; Paul Langford; Peter D. G. Thomas; John Brewer; H. T. Dickinson." Review by: R. Porter, *Past & Present*, 118(1), pp. 186–205.

Raver, A., 1996. "Cuttings: a garden with 40 years of character." *The New York Times*, 12 May. www.nytimes.com/1996/05/12/style/cuttings-a-garden-with-40-years-of-character.html [accessed 7 April 2014].

Roland, R. A., 1982. *Translating World Affairs*. Jefferson: McFarland.

Rüsen, J., 2005. "The visibility of history: bridging the gap between historiography and the fine arts." *Historein*, 5, pp. 130–41.

Sawyer, D. B., 2008. "Interpreter training and education at the US Department of State: origins, developments, and current practices." *Interpreting and Translation Studies – The Journal of the Japan Association for Interpreting and Translation Studies*, 8, pp. 267–77.

Thayer, C. W., 1959. *Diplomat*. New York: Harper.

Venuti, L., 1995. *The Translator's Invisibility: A History of Translation*. London: Routledge.

15

INTERPRETERS IN THE FIELD

Friends or foes?

María Manuela Fernández-Sánchez

In research on the history of interpreting, it is not a frequent occurrence to come across photographs of wartime interpreters, and even less frequent for the interpreters to be their main focus. Similarly to the wide range of other people and occupations (soldiers, doctors, nurses, and so on) depicted in war photography, whose portrayal is generally constrained by the cultural preconceptions and social stereotypes of the time, interpreters usually appear in a role-coded position, either near or between military officers and the person or people being interrogated. This position reflects both the function of interpreters as mediators and the marginality of their role (Bochner, 1981, p. 32). More interestingly, the visual character of the photographic source gives us access to the physical appearance of interpreters and their communicative activities, contributing to the documentation of an important dimension of armed conflicts that often goes unnoticed.

This study makes use of the descriptive information about interpreters provided by two photographs from the Western wars in Korea and Vietnam. Questions such as racial conflict tend to be censored in wartime, so the physical evidence that many photographs offer regarding Asians as interpreters in the particular conflicts examined here may be very eloquent testimony in itself. Moreover, by analysing these pictures we can address both the "preferred reading" or stories suggested by the caption (Hamilton, 1989, p. 175) and the technical and visual cues of the photographic images, as a way of understanding how photography was practised in wartime contexts.

Much of the recent work on war and photography is a rich conversation about the expectations, uses and limits of photographic realism in the digital era. In this context of new technologies and global communication, photographer and professor Fred Ritchin asks the following question: "How … , after all that is happening in computer imaging, can one safeguard the integrity of the photograph in its populist role as societal informant?" (1999, p. 72). The traditional role of journalistic photography as a visual record of war events, which was to be an instrument to inform the general public and to "awaken concern" (Berger, 1980, p. 43), experienced its golden days during the Vietnam War. This was "when war photographers were regarded as defenders of the truth and their photographs had the power to move opinion" (Struk, 2011, p. 5). Nevertheless, news agencies also imposed a certain degree of censorship and systematically refused to publish photos that were considered too shocking to use (Knightley, 2003, p. 449).

Long before Vietnam and Korea, World Wars I and II, the Spanish Civil War and even the American Civil War, photographers were already being sent – just a few decades after the invention of the camera during the first half of the nineteenth century – to cover early armed conflicts such as the Crimean War (1853–56). The objective was to obtain the realistic images afforded by the new medium. From the very beginning, photographic images were used to justify war and/or to balance out more critical reporting by press correspondents. For example, photographer Roger Fenton was encouraged to travel to Crimea and given a very specific mission: "Fenton's photographs, technically excellent though they are, portray a war where everything looks ship-shape and everyone happy" (Knightley, 2003, p. 14).

In wartime, governments place great importance on pictorial testimony. The control and use of these images, depending on the evolution of the conflict and public support for the war effort, make up a key element in any propaganda campaign. Accordingly, a high percentage of war photographs show carefully prepared scenes in which the participants are posed for propaganda purposes (for example Fenton's photographs of the Crimean War). In the opinion of Caroline Brothers, propaganda images "speak directly to the cultural concerns of the society at which they are directed, both in the subject chosen for representation and in the way those subjects are portrayed" (1997, p. 2).

When examining war photography, it is important to bear in mind the interpretive challenges that public images of this genre pose to present-day viewers. Our own historical distance from past conflicts, and the impressive tradition of Western war photography, dissuade us from going beyond iconic war photographs and reassuring war movies that always end in victory. As a result, "the reality of the war has more or less passed into film or dream, seemingly devoid of politics" (Walsh and Aulich, 1989, p. 1). The analysis of war photographs must thus include consideration of many signifying elements that explain the endurance of public images and their role in the construction of public memories. In the same way, it is also necessary to account for the broader process of the production and distribution of images in wartime, the different media involved, their more or less intrusive role, and their use in the service of institutional control. An additional dimension is the contrast between photographs as contemporary documents and as retrospective testimonies. Photographs speak to their own time and record contemporary events, particularly those regarded as newsworthy. Although they pursue different goals in their portrayals of people and topics, they usually all reflect social concerns. That is one of the reasons why documentary photography requires words and captions. At the same time, photographic images are the result of the technical advances and aesthetic conventions of the era (Sayer, 2010, p. 60). For instance, the practice of photojournalism expanded with the invention of hand-held cameras like the Leica, the growth of news agencies and the popularity of illustrated news magazines such as *Life*, founded in 1936.

To study photographic images of interpreters in settings of armed conflict is to acknowledge explicitly the presence of interpreters serving in wars. In doing so, we consider their contribution to the war effort. This not only fills a gap in historical research on the field, but also sheds light on the values associated with their visual representation as military interpreters in various armed conflicts, normally in line with their military rank. Drawing on recent methodological approaches to visual content analysis (van Leeuwen and Jewitt, 2001), I focus on a set of meaningful visual cues and the historical context within which the photograph can be decoded and understood.

Framing interpreters in the field

The two photographs studied here were taken in the first decades of the historical period known as the Cold War (1947–91): in the early stages of the Korean War (1950–53) and in the decade of active US combat involvement during the Vietnam War (1959–75). What distinguishes these pictures from others generated by military history, which also depict military interpreters at work interrogating civilians or prisoners of war, is the unusual centrality of the interpreter or of certain salient attributes of the interpreter's work in the two pictures' composition.

Figure 15.1 shows an Asian woman, a Chinese interpreter, who is being interviewed by UN correspondents. In Figure 15.2, the interpreter is alone, facing the

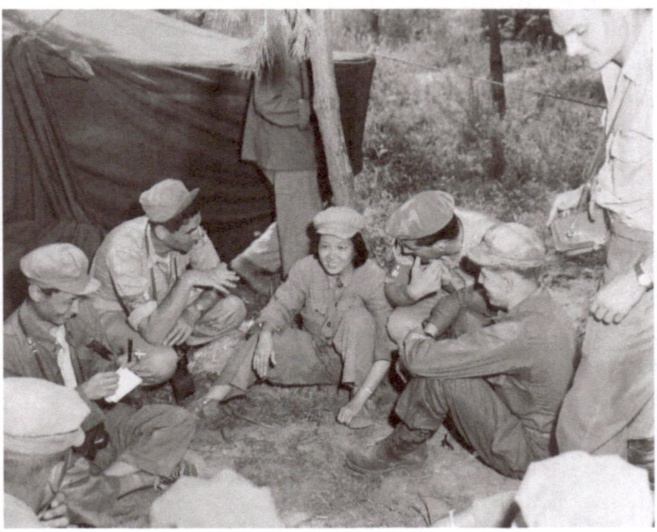

FIGURE 15.1 UN correspondents interview Private Chang, a woman interpreter for the Chinese People's Army

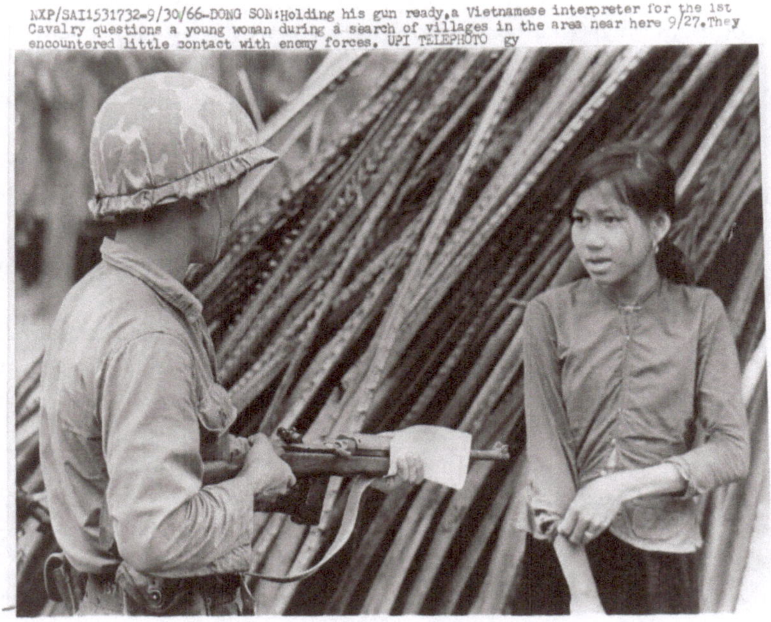

FIGURE 15.2 First Cavalry questions young Vietnamese

"enemy". He is armed, and looks more like a soldier than a linguistic facilitator and mediator. The two interpreters are speakers of Chinese and Vietnamese respectively, which makes sense in these historical contexts.

Framing interpreters in Korea: A sanitized version of the war

Some scholars have observed that the Korean War is one of the least-represented wars in US cinema (Muraire, 2008, p. 121). Comparing combat images from the Korean and Vietnamese conflicts, other authors underline the "more sanitized version of the combat experience" portrayed in Korean War photographs (Sherer, 1988, p. 755). This cultural representation and coverage of the conflict are partly explained by the confidence that pervaded the public culture of the 1950s (Whitfield, 1996, p. 231) and the political atmosphere in the United States, where "not one major daily newspaper opposed the war" (Knightley, 2003, p. 379).

With respect to photographic images depicting interpreters in the Korean War, the sources consulted document the presence of interpreters from different geographical and cultural backgrounds in a variety of settings. The most significant of these was the negotiating process leading up to the signing of the Korean Armistice Agreement, which was conducted between US military officers, representing the UN Command, and Chinese and North Korean military personnel. The truce talks began in the Communist-controlled town of Kaesong on 10 July 1951, moved to the more neutral town of Panmunjom on 25 October 1951, and achieved the signing of the armistice agreement only on 27 July 1953. In the United States, the war in Korea has often been considered a foreign war, since it was fought under the UN flag, but there is no question that the Korean conflict was deeply disturbing to the American public for many reasons, including the frustrating truce talks and the unclear victory.

The scene captured in the first photograph is from that major event in the course of the Korean War, the truce talks. The caption reads: "UN correspondents Alfred van Sprang of Holland (second from left) and Alain de Prelle of Belgium (fourth from left) interview Private Chang, a woman interpreter for the Chinese People's Army, at the armistice conference site in Kaesong, Korea". The author of the photograph is Donald Hildreth (Private First Class, 51st Signals Battalion). The black-and-white photograph is one of a few, also taken by Private Hildreth, that depict press correspondents covering the armistice negotiations. An initial conclusion regarding the series of photographs on the same subject is that they were motivated by the photographer's desire to document the activities of these journalists. The idea of telling "parallel" stories in war situations goes hand in hand with the role of the photographer as an eyewitness of war. This sense of opportunity has been intrinsic to the practice of documentary photography since the beginnings of the art in the first half of the twentieth century, and is associated with such names as Robert Capa.

The photo portrays a relaxed group of equals of different nationalities and ethnic backgrounds. There are seven white men, an Asian man and an Asian woman. Three of the white men are seen from the back and only parts of their caps are visible. At the front right of the photo, a man is standing with a camera around his neck. The background shows a military tent and a man walking by. The other men are in a circle around the Asian woman, who is the main focus in the scene. Two men, squatting down and bending over her, are asking her questions and seem to be very interested in her answers. The rest of the people are sitting on the ground and might also be war correspondents.

According to the caption as set out in the Imperial War Museum archive, the two men closest to the woman are UN correspondents, and the woman is an interpreter for the Chinese People's Army. The UN correspondents wear caps and comfortable summer uniforms. The female interpreter, who has glasses, is wearing the typical drab military uniform of Communist officers. The Asian man, also dressed in a plain uniform, is folding a sheet of paper and holding a pen. The fact that the people seem to be enjoying the conversation is reflected in their relaxed body language. This is particularly the case of the woman. Almost all the figures, including the interpreter, are smiling, which suggests that they have a familiar relationship or at least have spent some time together at the armistice conference site in Kaesong.

The general impression is that the picture has been carefully stage-managed to make it more interesting for US audiences. This practice of setting scenes is common in photojournalism (Struk, 2011, p. 5). The detached position apparently sought by the photographer is achieved thanks to the combination of a number of elements, such as the camera perspective from above and at least two elements that might detract from the artistic composition of the photo: a man walking by (whose head is not visible, only his body), and the reporter who is sliding into the photo from the right side of the frame. The aesthetic impact of the photograph lies in the carefully arranged composition of the group and its symmetry, with the interpreter and the tree behind her, bisecting the picture. The normal view and the frontal position from which the photograph is taken underline the centrality of the interpreter and the people close to her.

At the time when the Korean War broke out, photography was very popular both on a professional and non-professional level. The conflict was mainly covered by journalists working for news agencies and illustrated news magazines. In contrast with the "living-room war" of the conflict in Vietnam, during the Korean War television reporting was still in its infancy. From the beginning of the war, news reporters from the United States and UN allied countries worked in Korea. They were men and also women of various nationalities, who moved with the armed forces to cover the conflict. During the first year of the war, no restrictions were placed on the information broadcast or reported, only a voluntary code of war aimed at preserving military secrecy (Knightley, 2003, p. 367). However, as the war progressed, US military commanders began to censor the news to a greater or lesser degree, and during the truce talks this censorship reached its height. For this reason, Western war correspondents, who needed to find news to report, began to consult Communist sources as well as the Western correspondents working for the North Korean-Chinese delegation (pp. 386–87). In fact, it may well be that the photographer took the picture precisely with the purpose of publicizing this strategy and the constraints that had led to it. In this sense, the picture could be regarded as one more testimony of the problematic nature of the relationship between war correspondents and military authorities in the first stage of truce talks in Korea. However, the feeling of comradeship captured in the photograph, the close proximity of people belonging to different nationalities, ideologies, social levels, race and gender in the hostile military context of the armistice meeting also gave the photographer a unique opportunity to convey a reassuring message of security, normality and eventual peace to the US audience.

Framing interpreters in Vietnam: Arresting photographs

In his chapter "Photographs of Agony", John Berger explains the literal meaning of the adjective "arresting" as applied to dramatic war photographs. He says: "We are seized by them. … As we look at them, the moment of the other's suffering engulfs us" (1980, p. 42). The Vietnam War has the dubious distinction of having produced some of the most powerful war images. The Vietnam War "changed photojournalists' function", writes Ritchin (1999, p. 36). In a significant way, photographers and television reporters informed the US public about the drama in Vietnam much better and faster than their editors and the government authorities, although their work was constrained and shaped by US government propaganda (Appy, 2003, p. 238). Former anti-war activist Noam Chomsky observes that the My Lai massacre became a big issue only in November 1969, a year and a half after the killings took place (Mitchell and Schoeffel, 2002, p. 35). Studying this complex propaganda framework, Michael Sherer shows that, when public support for the war in Vietnam was strongest, from January 1965 to July 1967, the most important magazines published images mostly of US forces and military equipment working in combat-related and non-combat scenes. As public support of the war waned, the type of photograph also changed, to portray combat situations with less focus on US forces and equipment (Sherer, 1989, p. 394).

The second photograph (Figure 15.2) is from United Press International. It is a wire photograph or telephoto taken by an anonymous photographer. The caption reads: "DONG SON: Holding his gun ready, a Vietnamese interpreter for the 1st Cavalry questions a young woman during a search of villages in the area near here 9/27. They encountered little contact with enemy forces." The picture was taken on 27 September 1966, probably in rural South Vietnam. In 1966 and 1967, the Vietnam War had entered the second phase of the plan conceived by General William Westmoreland. During that period, "search and destroy" missions combined US ground troops with supporting artillery and airpower. In these missions, US and South Vietnamese soldiers patrolled the countryside in search of Viet Cong guerrillas and their military installations. As the caption suggests, most of the time soldiers were not fighting, but interrogating civilians in small hamlets, normally women and old people.

The photograph shows a South Vietnamese military interpreter on patrol. He is armed and points his weapon at a young unarmed Vietnamese woman. The image shows a side-shot of the interpreter. He is wearing a uniform and combat cap, and carries a standard firearm in Vietnam War, a semi-automatic carbine. His left hand grasps the lower part of the weapon and at the same time holds a sheet of paper. In fact, the piece of white paper is the central element in the composition of the photograph, the *punctum* in Barthes's terms (1984) – the unexpected element that highlights the strangeness of the relationship between one of the most conventional instruments used by interpreters and a weapon, normally associated with the military. The young woman is close to the interpreter. She is dressed like a Vietnamese peasant and looks at him with suspicion. Even though she may be frightened, she still maintains eye contact with the interpreter. Her body language, however, is more relaxed than her face. Both are South Vietnamese, and appear to be very young.

The photo, which is in black and white, is quite disturbing, a typical feature of war photography. Hamilton (1989, p. 174) observes that rarity and drama are the conditions considered newsworthy for war photographs. Looking at this photograph we are troubled by the tension of the military patrol context, but also by the dramatic multi-level contrast between the two people portrayed: military/civilian, armed/unarmed, and man/woman. The emotional impact of the subject is enhanced by the professional work of the photographer. For example, the side-shot of the interpreter is a close-up. The central focus of the picture is the piece of white paper held by the military interpreter. The photograph is taken from a frontal position, which signifies greater involvement both with the young woman and with the viewer being invited to look at the scene.

Two final points can be made here. The display of military control is effective, but not excessive: the interpreter is armed with a light carbine, an inferior weapon compared to the newly developed M-16 rifle, "the American answer to the enemy's AK-47" (Westmoreland 1976, p. 158). Secondly, the drama of the scene is mitigated because it is not a life-threatening situation. The combination of these two elements makes this picture an extraordinary example of the social concerns and demands that photography was supposed to meet in the initial stages of the Vietnam War.

Conclusions

Photographs depicting interpreters serving in war zones raise many questions about the presence of interpreters as photographic subjects and about the projection of particular values – values added by these images – onto the cultural representation of specific armed conflicts. The photographs analysed in this chapter were taken during the Korean and Vietnam Wars and are good examples of the practice of photojournalism at those times. Both pictures tell a story, a military story, about the war and reflect the photographer's agenda of realistically portraying the war, which was consistent with his or her still unquestioned role as an eyewitness of the war being covered. From this perspective, the instrumental function of the interpreters portrayed in both pictures, as valuable sources of information and as military linguistic facilitators, contributed in each case to furthering US propaganda objectives at these stages of the conflicts. However, as suggested by the title of this study, the status of the military interpreters as friends or foes is ambiguous. In fact, the unusual centrality of the interpreters in the photos stands in vivid contrast with their marginal role in the military context of the conflicts being portrayed, providing an invaluable insight into the military culture of the time. Racial and gender prejudice seems to disappear when parallel stories are sought, whereas the markedly racist and violent nature of the war comes to light when a South Vietnamese interpreter is used in the field to patrol enemy territory and at the same time to interrogate a young Vietnamese girl.

References

Appy, C. G., 2003. *Patriots: The Vietnam War Remembered from all Sides*. New York: Viking.

Barthes, R., 1984. *Camera Lucida: Reflections on Photography*. Translated by R. Howard. London: Fontana.

Berger, J., 1980. *About Looking*. London: Bloomsbury.

Bochner, S., 1981. "The social psychology of cultural mediation." In: S. Bochner, ed. *The Mediating Person: Bridges Between Cultures*. Cambridge, MA: Schenkman. pp. 6–36.

Brothers, C., 1997. *War and Photography: A Cultural History*. London: Routledge.

Hamilton, R., 1989. "Image and context: the production and reproduction of *The Execution of a VC Suspect* by Eddie Adams." In: J. Walsh and J. Aulich, eds. *Vietnam Images: War and Representation*. New York: St. Martin's Press. pp. 171–83.

Knightley, P., 2003. *The First Casualty: The War Correspondent as Hero, Propagandist and Myth-maker from the Crimea to Iraq*. London: André Deutsch.
Mitchell, P. R. and Schoeffel, J., 2002. *Understanding Power: The Indispensable Chomsky*. New York: The New Press.
Muraire, A., 2008. "Notes on the American war film from the forties to the eighties." In: G. Hugues and K. Hildenbrand, eds. *Images of War and War of Images*. Newcastle: Cambridge Scholars. pp. 121–41.
Ritchin, F., 1999. *In Our Own Image: The Coming Revolution in Photography*. New York: Aperture.
Sayer, D., 2010. "The photograph: the still image." In: S. Barber and C. M. Peniston-Bird, eds. *History Beyond the Text: A Student's Guide to Approaching Alternative Sources*. New York: Routledge. pp. 49–71.
Sherer, M., 1988. "Comparing magazine photos of Vietnam and Korean wars." *Journalism Quarterly*, 65(3), pp. 752–56.
Sherer, M. D., 1989. "Vietnam war photos and public opinion." *Journalism Quarterly*, 66(2), pp. 391–95, 530.
Struk, J., 2011. *Private Pictures: Soldiers' Inside View of War*. London: I.B. Tauris.
van Leeuwen, T. and Jewitt, C., 2001. *Handbook of Visual Analysis*. London: Sage.
Walsh, J. and Aulich, J., 1989. "Introduction." In: J. Walsh and J. Aulich, eds. *Vietnam Images: War and Representation*. New York: St. Martin's Press. pp. 1–9.
Westmoreland, W. C., 1976. *A Soldier Reports*. Garden City, NY: Doubleday.
Whitfield, S. J., 1996. *The Culture of the Cold War*. Baltimore, MD: The Johns Hopkins University Press.

16

"THE BIGGEST ROUND TABLE"

The interpreters' visibility at the Potsdam Conference

Icíar Alonso-Araguás

Finding visual materials as primary sources to document interpreting as a social practice may be difficult for some historical periods or settings, but wartime settings are sometimes an exception. This chapter discusses the presence of diplomatic interpreters in a photograph dated 1945, a highly sensitive period in contemporary history and in the history of interpreting. Based on several sources, such as photographic records, memoirs, institutional archives and the contemporary press, I explore the material dimension of what has been traditionally called the interpreter's invisibility in historical events.

Two preliminary methodological remarks may be useful. The first concerns the nature of photographs as historical sources and how they can cast light on the representation and perception of the interpreter. Images, and photographs in particular, are not in themselves historical sources but a way of recreating the world. Only when we, as researchers, look at them from our own perspective do photographs become historical sources with many possible meanings (De las Heras, 2012, p. 24). This is equally true of institutional records, which are also subject to interpretation. As Reynolds notes (2008, p. 9), "they enable us to construct a reality that no participant could have known at the time because he could not see the other man's cards". Images enable greater insight into extralinguistic aspects such as body language, protocol issues and spatial arrangements. Even if different filters will be at work on all of these sources (De las Heras, 2012, pp. 34–35), a critical approach that combines several such sources may lead to a more comprehensive understanding of the topic under discussion: how interpreters are highlighted by the photographs.

The second methodological question is how interpreters should be understood as historical subjects. In this context, I regard interpreting as a socially localized practice (Inghilleri, 2003; Wolf, 2007) and interpreters as social agents, moving through history and naturally embedded within a social frame. Such a frame is associated with a complex network of juxtaposed spaces, where other stories are happening at the same time. To this extent, it is useful to think in terms of Foucault's sociology of space (1986) and to apply some of his suggestions to this case study by looking at the photograph as an "other space" or *heterotopia*. That means going beyond the description of the framed event and exploring different levels of meaning through its juxtaposed spaces – the micro-space within the picture and the network of spaces outside the picture – even if such a manipulation might be considered an anachronism. Of course, our particular localization in the present-day world differs from that of the 1940s audience. As Foucault (1972, p. 5) also argues, "historical descriptions are necessarily ordered by the present state of knowledge, they increase with every transformation and never cease, in turn, to break with themselves".

Large audiences are not necessarily familiar with interpreters, even when they are clearly visible through photographs, so captions become especially relevant in order to make sense of interpreting as a social practice. The interdependence between pictures and written discourse may also be reinforced by other historiographical genres, such as interpreters' memoirs. Apart from several more technical issues, Churchill's interpreter, for example, explained in detail the feelings that overwhelmed the interpreter in such highly charged and historic events:

> What happens inside him? Nothing happens, beyond a sinking feeling and a sense of loneliness. He is alone in the midst of an ocean of words. He just interprets. I imagine that a musician does the same: he just plays.
>
> *(Birse, 1967, p. 110)*

Looking at the photograph both as a documentary image and as a *heterotopia*, then, my intention is to slowly zoom in to different levels of intensity and explore the relationship between the representation of interpreters and the social practice that it reveals.

The cards on the table

The picture (Figure 16.1) belongs to the Truman Library (TL) and was taken on 25 July 1945 at the beginning of one of the Potsdam Conference's 13 plenary meetings. The first session took place on the afternoon of 17 July and the last ended on 1 August, after midnight. The complete caption reads:

> President Harry S. Truman, British Prime Minister Winston Churchill, and Soviet Union Prime Minister Josef Stalin and their advisors are seated around the round conference table prior to the opening of one of the Potsdam Conference meetings in Potsdam, Germany. President Truman has his back to the camera; Prime Minister Stalin is on the right side of the photograph; Prime Minister Churchill is on the left side; they are in the chairs with more ornamentation on the backs. Ambassador Joseph Davies and interpreter Charles Bohlen are seated to the left of President Truman in the photograph; Secretary of State James Byrnes and Admiral William Leahy are seated to the right of President Truman. From the album, number 2: "President Truman's Trip to Potsdam".

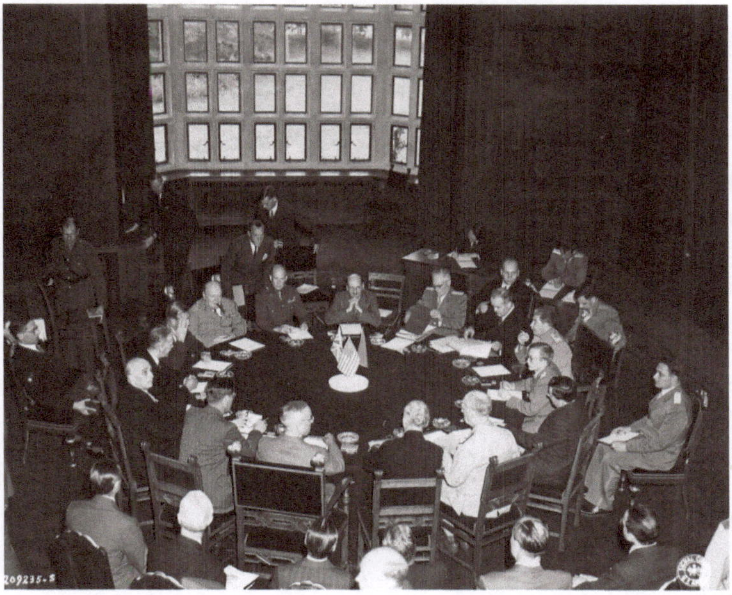

FIGURE 16.1 The Big Three with their advisors prior to opening of meetings of the Potsdam Conference

The photographer is unknown, although the US Army Signal Corps is mentioned as institutional originator. Photographers had access to the conference room only at the beginning of each meeting, as has been confirmed by the series of photographs in the TL and by one of the participants (Birse, 1967, p. 206).

The date of the photograph coincides with Truman's "First Hundred Days", and it also appeared in the 6 August issue of the US magazine *Life*, the day when the first atomic bomb was dropped on Hiroshima.

It is a black-and-white photograph, 8 x 10 inches (21 x 26 cm), with an almost full frame and a vertical shot. It has a clear documentary intention, but it is the photographer, in accordance with his own communicative strategy, who selects the framed space, choosing this fragment of reality and making everything outside the photograph non-existent. The high-angle shot he uses results in a sort of diagram or cartography of power. The framing is very strongly centred, and a symmetrical composition in concentric circles can be identified: a centrepiece with three small flags occupies the middle of the image and of the table, around which the other elements are displayed – the circle of ashtrays and documents, the first circle of chairs, and two more circles beyond that. The main source of natural light for the scene is the big window at the back, but we may assume that the room also had artificial lighting over the table.

The visual composition shows a clear contrast between round and rectilinear forms. The big, round table covered by a fine burgundy cloth – here in black and white – appears in the foreground and it contrasts with the huge rectangular glass window in the background. The table monopolizes the space, although both the room and the building, a 176-room palace, were as monumental in size as the furniture:

> It was a fantastic house, with numerous pseudo-Gothic spires, curious angles, castellated walls, and stained glass windows. … An immense reception room which occupied most of the ground floor was used as the Conference Room, and there were smaller rooms adjoining, each of the three Delegations being allotted one of these as ante-rooms and offices. The Conference Room contained the biggest round table I had until then seen.
>
> *(Birse, 1967, p. 206)*

The disposition of the concentric circles also shows a hierarchy of power among the persons represented in the picture. In order to preserve both the physical and the symbolic appearance of a supposed balance, protocol rules required heads of delegation to make their entrances and exits through separate doors of precisely the same size (Dobbs, 2012, p. 298).

The stage

The photograph provides a general view of the conference table. The 15 main actors – the Big Three, each flanked by his interpreter and three advisers – and at least six assistants from each delegation wait for the proceedings to start. The space occupied by each member was decided in advance, and three decorated armchairs were reserved for the principals. As the characters in a drama, all the participants seem to be ready to start when the chairman, Truman – following Stalin's proposal – calls the meeting to order. The static nature of the scene is slightly altered by the centrifugal force represented by the attendants at the edge of the table.

The official description of the picture only identifies the members of the US delegation, in addition to Churchill and Stalin. Around the table, clockwise from Truman, are Churchill (replaced from 26 July by Clement Attlee), on the left, and Stalin on the right. Grouped around Churchill are Attlee, Eden, Sir A. Cadogan and the interpreter Birse; grouped around Stalin are Molotov, Vyshinsky, Gromyko and the interpreter Pavlov; around Truman are Ambassador Davies and the interpreter Bohlen, on his left, and Byrne and Leahy on his right. The main participants were almost the same ones as those present at Yalta, the only difference being that Roosevelt had passed away in April 1945 and was replaced by Truman. Furthermore, during meetings other than plenary sessions, experts filled the room and interpreters had to raise their voices above the rustle of papers, loud whispers, coughing and shuffling of feet (Birse, 1967, p. 206), so noise would have been an additional challenge for interpreters.

It is worth noting that the interpreters are seated at the table, to the left of their principals rather than behind them. They are clearly visible and take up a similar hierarchical position to that of the other participants, as had been the case for diplomatic interpreters in the first decades of the twentieth century – the period of splendour for consecutive interpreting in international conferences – before the introduction of simultaneous interpreting.[1] Birse and Pavlov are wearing military uniforms, like their principals. Birse held the rank of Major in the British military and Pavlov had been obliged by Stalin to wear a uniform as well. The image shows only male figures, at a time when the conference interpreting profession had not undergone the feminization process that can be traced by observing the interpreters' staffing tables since the early days of the United Nations.

The shape of the negotiating table, a round table that avoids hierarchies, is an icon of national status and sharing between the participants, and symbolizes the summit conference as a diplomatic tool (Goldstein, 1996, p. 27). This "diplomacy by conference" differs from diplomacy based on secret negotiations in the corridors of embassies or ministries. It has a clear visual component, a *mise en scène*, addressed to public opinion at large and – like any other communication commodity – usually covered by the big newspapers and wire agencies (Thayer, 1959, p. 107).

Inside the room, beyond the room

The ample window in the back wall serves as a light source but also as a link with the outside world, the outskirts of Berlin. The meeting took place at the Cecilienhof, Wilhelm II's palace and a symbolic site of the Reich. Potsdam was the last of a series of conferences between the Big Three Allies, after Tehran (1943) and Yalta (1945). But the conference, code-named "Terminal", also marked a break-up of the political status quo of the wartime alliance and the start of the nuclear age (Holmes, 2000, p. 38). From then on, the relations between the Allies would be less than friendly, opening the era of the Cold War. Thus, the picture reflects a highly critical period between "two very different wars and two very different worlds" (Dobbs, 2012, p. vii).

Based on the press images, the atmosphere in the plenary sessions appeared extremely courteous, and the polite gestures and warm greetings outside the room sent an image of a relaxed cordiality. Yet, only a few months later, this rhetoric and diplomacy were replaced by an "iron curtain" already presaged by the disagreements during the Conference:

> Churchill ... said of the status of the British and American representatives in the Soviet-controlled countries that "an iron curtain has come down around them". Stalin replied that this was "all fairy tales". Molotov and Eden both became angry during the debates.
>
> *(Bohlen, 1973, p. 234)*

However, the main issue hanging over the table on that 25 July was not discussed directly there, but in the corridors and ante-rooms during the breaks. In the days prior to this photograph, Truman was informed of the United States' first successful atomic bomb test and after discussion with Churchill he decided to tell Stalin about it in the most informal manner possible. This happened on the afternoon of

24 July, after the plenary session. Instead of communicating the information through his usual interpreter, Bohlen, he did so through Pavlov, the Russian interpreter, thus creating the impression that the exchange was not important: "I did not hear the conversation, although Truman and Byrnes both reported that I was there", notes Bohlen (1973, p. 237). However, the third interpreter states that both Bohlen and Pavlov were interpreting at that conversation (Birse, 1967, p. 210). In his log entry of 25 July, Truman's handwritten notes state:

> We met at 11 A.M. today. That is Stalin, Churchill, and the U.S. President. But I had a most important session with Lord Mountbatten and General Marshall before that. We have discovered the most terrible bomb in the history of the world. ... This weapon is to be used against Japan between now and August 10th.
>
> *(Truman Library. Papers 1945)*

The most immediate consequence of this meeting took place the following day, after Churchill and Attlee had left the conference and returned to London. It was the issuance of the Potsdam Declaration urging Japan to surrender unconditionally. The Japanese Prime Minister replied on 28 July at a press conference where he supposedly said: "We simply *mokusatsu-suru*. The only alternative for us is to be determined to continue our fight till the end" (Torikai, 2009, p. 33). This reply would cause the controversial translation mistake that may have been behind the version conveyed by Western news agencies, reporting the "refusal" (instead of the disregard, or "not worthy of comment") by Japan of the Allied ultimatum.[2]

The three interpreters

The interpreters for the Big Three at Potsdam were Charles E. Bohlen, Vladimir N. Pavlov and Arthur H. Birse.[3] Although they played the same part and sat in identical seats, they are identified in the photograph caption published by *Life* magazine as "assistant to the Secretary of State", "interpreter" and "translator" respectively, three different terms for one and the same role. However, the official caption from the TL does describe Bohlen as President Truman's interpreter. The three men had previously met while working in Tehran and Yalta, and Pavlov and Birse were on duty together at most of the bilateral meetings between Stalin and Churchill.

Interpreters' memoirs offer an alternative way to go beyond institutional records and cross-reference them with other kinds of narratives about the inner perceptions of the interpreter. Bohlen and Birse published memoirs which refer in detail to the event depicted in our photograph. Bohlen's approach focuses on the diplomatic activity, while Birse prioritizes the interpreter's perspective, paying attention not only to the historic events but also to his professional performance and the technical challenges he encountered. Pavlov refused to write his memoirs and tried to avoid contact with journalists, so the available references to him are indirect, mainly from his wife and colleagues.

Charles E. Bohlen (1904–74), a career member of the Foreign Service, was the personal interpreter for President Roosevelt in Tehran and Yalta and for President Truman at Potsdam. He also acted as a high-level interpreter in many other international meetings. Thanks to his profound knowledge of Soviet affairs, he belonged to the closest circle of presidential advisers and participated as an expert in US foreign policy strategy. For this reason, he held several posts of high responsibility in the US Administration, including ambassador (Truman Library n.d. Profile of Charles E. Bohlen). In May 2006 he was honoured with his own US postage stamp.

According to his memoirs, Bohlen's main area of interest was the diplomatic service, although his presence in Potsdam was also justified by his interpreting skills. This privileged overlapping made him an exemplary case of what it takes to become a diplomatic interpreter, similarly to interpreters such as Andrew Dickson White or Vernon Walters (Delisle and Woodsworth, 1995, pp. 271–72). When Truman came to power, Bohlen's relations with the President's inner circle were not as close ("I was not present at most of the intimate conversations between Truman and Byrnes") and he explicitly says that interpreting was regrettably his main activity at Potsdam (Bohlen, 1973, p. 228).

Vladimir N. Pavlov (1915–93) was the personal English-speaking interpreter for Stalin and Molotov, and his career was directly linked to both of these in the most literal sense:

> I had never thought of becoming an interpreter. … But I liked languages and I was good at them. … I did not want to be a diplomat. I was twenty-four and my only interest was to become a scientist. I told Molotov this honestly in our first conversation, but he interrupted me abruptly: "You are a communist and you will work where you are needed most."
>
> *(Torchinov and Leontyuk, 2000)*

After obtaining his engineering degree, he was called to the Communist Party headquarters to take a test in German and, in view of the excellent results, the authorities forced him to become an interpreter. His voluntary retreat from the diplomatic stage immediately after the dictator's death seems a clear indication that he was not happy carrying out those duties. Despite his avoidance of the media, Pavlov is an astonishing example of visibility. According to his widow (Gridneva, 1999), Stalin obliged him to take front stage in many photographs, even when he was not interpreting. One reliable source of information regarding Pavlov is his British colleague Birse:

> His English grammar may not have been perfect, but he was seldom at a loss for a word, and could adjust his speed, emphasis, and tone accordingly. We worked as a pair almost uninterruptedly for three years. His presence gave me confidence.
>
> *(Birse, 1967, p. 113)*

Arthur Birse (1889–1981) was a high-level interpreter and Major in the British Army. His memoirs narrate historic events and international meetings during the twentieth century in remarkable detail. They are an endless source of information on diplomatic interpreting and, from a wider perspective, on the business of the interpreter. Birse was the son of a Scottish merchant established in St Petersburg and he attended Russian schools until the Bolshevik revolution. He then joined the British military legation as a translator and ended up settling in Britain, where he worked in the banking sector. In 1940 he joined the Intelligence Department of the British military. There he started accompanying higher-ranking officials throughout Europe as a linguistic assistant and from 1941 also worked in the USSR at British diplomatic and military missions. His excellent command of languages and extensive knowledge of different cultures and Soviet foreign policy allowed him to gain hands-on interpreting experience in the field.

Conclusion

The network of spaces suggested by this photograph offers two major clues regarding the traditionally controversial issue of interpreters' invisibility at historic events, which seems to be transformed into different levels of visibility in critical periods when key actors are involved, as at the Potsdam round table.

First, in this photograph the three interpreters' visibility and socio-professional recognition are enhanced in several ways. They occupy a similar position to their diplomatic delegations, they follow the same dress code (military or civil) and they carry out additional advisory functions on military, diplomatic or cultural issues.

Second, by assimilating them into the group of their colleagues, the interpreters' enhanced social recognition paradoxically also makes them invisible to the public – the potential addressee of the photograph – and to researchers. The image requires additional visual and written discourse, not only captions, in order to convey more meaningful information on the interpreters' identities and roles and on interpreting as a social practice.

As a historical record, then, the picture points to a juxtaposition of several different spaces – physical, chronological, symbolic or social – inside and beyond this single real place (Foucault 1986), the Potsdam round table: the materiality of the room, separated from the space outside; the time that runs parallel to many other "contemporaneities"; the symbolic representation of a microcosmic geopolitical cocoon at once global and local; and the social dimension found in the interconnections of actors with different roles and backgrounds.

Notes

1 For further details on this period, see Baigorri-Jalón (2000). Simultaneous interpreting, born in the late 1920s at the International Labour Organization, was tested successfully in 1945–46 at Nuremberg, and a few months later at the UN. Although it was adopted as the preferred system by the UN, it was not used at other international summits. According to Reynolds (2008, p. 335), the United States did not test it for the highest-level

bilateral meetings until 1981, and until 1985 for the plenary meetings in Soviet-American summits.

2 As Torikai notes (2009, p. 33), the choice of such an ambiguous and culturally embedded word to answer the ultimatum could be seen as a lack of awareness of the requirements of translation in cross-cultural communication.

3 They were not the only interpreters present. Birse (1967, p. 206) reported that among the assistants of the British delegation was "my friend Theakstone, who had come from Moscow to help with the interpreting". See also references to the British interpreter H. Lunghi in Smith (2005).

References

Baigorri-Jalón, J., 2000. *La interpretación de conferencias: el nacimiento de una profesión. De Paris a Nuremberg* [*From Paris to Nuremberg: The Birth of Conference Interpreting*]. Granada: Comares.

Birse, A. H., 1967. *Memoirs of an Interpreter*. New York: Coward-McCann.

Bohlen, C. E., 1973. *Witness to History 1929–1969*. New York: Norton.

De las Heras, B., 2012. *El testimonio de las imágenes: fotografía e historia* [*The Testimony of Images: Photography and History*]. Madrid: CVG.

Delisle, J. and Woodsworth, J. eds., 1995. *Translators Through History*. Amsterdam: John Benjamins.

Dobbs, M., 2012. *Six Months in 1945. FDR, Stalin, Churchill, and Truman – from World War to Cold War*. London: Hutchinson.

Foucault, M., 1972. *The Archaeology of Knowledge*. Translated by A. M. Sheridan Smith. London: Tavistock.

——, 1986. "Of other spaces." Translated by J. Miskowiec. *Diacritics*, 16(1), pp. 22–27.

Goldstein, E., 1996. "The origins of summit diplomacy." In: D. H. Dunn, ed. *Diplomacy at the Highest Level: The Evolution of International Summitry*. London: Macmillan. pp. 23–37.

Gridneva, N., 1999. "Perevódchikom Stálina bil inyenér-teplotéjnik" ["Stalin's interpreter was an engineer of thermal energy"]. *Komersant*. http://kommersant.ru/doc/214282 [accessed 7 April 2014].

Holmes, R., 2000. *World War II in Photography*. London: Carlton.

Inghilleri, M., 2003. "Habitus, field and discourse: interpreting as a socially situated activity." *Target*, 15(2), pp. 243–68.

Reynolds, D., 2008. *Summits: Six Meetings that Shaped the Twentieth Century*. London: Penguin.

Smith, D., 2005. "There was a heap of ashes. Someone said it was Hitler": the first Briton in the Berlin bunker recalls that day – and meeting Stalin. *The Observer*, 1 May. www.guardian.co.uk/uk/2005/may/01/artsandhumanities.secondworldwar [accessed 7 April 2014].

Thayer, C. W., 1959. *Diplomat*. New York: Harper & Brothers.

Torchinov, V. A. and Leontyuk, A. M., 2000. *Vokrug Stalina, Istóriko-biografícheskii správochnik* [*About Stalin: Historical-biographical Guide*]. St. Petersburg. http://hrono.ru/biograf/bio_p/pavlov_vn.html [accessed 7 April 2014].

Torikai, K., 2009. *Voices of the Invisible Presence: Diplomatic Interpreters in Post-World War II Japan*. Amsterdam: John Benjamins.

Truman Library. 1945. "Papers of Harry S. Truman. Diary, 17 July, 18 July, 25 July." www.trumanlibrary.org/whistlestop/study_collections/bomb/large/index.php [accessed 7 April 2014].

——. n.d. "Profile of Charles E. Bohlen." www.trumanlibrary.org/profile/viewpro.php?pid=279 [accessed 7 April 2014].

Truman Library Photographs. www.trumanlibrary.org/photographs [accessed 7 April 2014].

Wolf, M., 2007. "The emergence of a sociology of translation." In: M. Wolf and A. Fukari, eds. *Constructing a Sociology of Translation*. Amsterdam: John Benjamins. pp. 1–36.

17

THROUGH THE COLD WAR LENS

Russian and US interpreters as cultural and political mediators

Brian James Baer

> Richards and Ogden in their *Meaning of Meaning* point out that translation is not always as reliable as people think (and it is one of the incident purposes of this paper to show that it is still less reliable than Ogden and Richards make it).
>
> (Edmond S. Glenn, 1955, p. 21)[1]

In analysing photographs of interpreters and interpreted communicative events of the early Cold War era, we bring together two cultural practices that were in many ways emblematic of the positivist view of representation that dominated the culture of the time.[2] Within positivist models, which assume that an anterior world "independent from language" exists and that our knowledge of that world can be "derived from sensorial data, which constitute a reliable basis for cognition" (Glenn, 1955, pp. 3–4), photography and interpreting were seen as privileged forms of mimesis, exemplars of a realist aesthetics described by Scott McQuire in *Visions of Modernity* as asserting "the formal priority of reality over its images, the 'thing itself' over the representation, [and] the precedence of original over copy" (1998, p. 17). To the extent that photography and interpreting were viewed as virtually "synonymous with fidelity in representation" (p. 13) – as one UN interpreter told a *New York Times* reporter in 1955: "You must be *absolutely faithful* to what the delegate is trying to convey" (White, 1955, p. 12; emphasis added) – photographers and interpreters were constructed in the discourse of the time, alongside scientists, as neutral conduits or faithful recorders of "the thing itself". It is no coincidence, then, that the Golden Age of photographic journalism and the professionalization of interpreting took place against the backdrop of the Space Race, as did the first major developments in the theory and practice of machine translation, which promulgated a view of translation as code-switching or linguistic matching (Raley, 2003).

The Cold War period was also, however, a time when a major theoretical assault on positivism was under way in linguistics and other fields in the humanities. Led by proponents of linguistic relativity, it was represented perhaps most influentially by the work of the anthropologists Edward Sapir and Benjamin Lee Whorf, who argued that language was not a representation of an anterior reality but rather played a role in constructing that reality. While they published their major work before the end of the Second World War, their ideas were debated throughout the 1950s, and the term "Sapir-Whorf hypothesis" was coined only in 1954, by Harry Hoijer.

In my analysis of several iconic Cold War photographs of interpreted communicative events and the captions accompanying them, I will show how the political pressures of the Cold War era challenged positivist pieties concerning fidelity and equivalence, initiating a debate, still ongoing today, about the role and agency of the interpreter.

The photographs

The three photographs that will be the subject of my analysis were all taken by US photojournalists and feature interpreters and interpreted communicative events that "capture" important moments in Cold War history and culture. Figure 17.1 portrays an iconic moment in Cold War history during the Soviet Premier Nikita Khrushchev's first and only visit to the United States, a 12-day tour that involved meetings with various political and business leaders. This first-ever visit by a Soviet premier to the United States occurred only a few months after Vice-President Richard M. Nixon's widely publicized visit to the USSR for the opening of the American National Exhibit in Moscow, where the now famous "kitchen debate" took place.

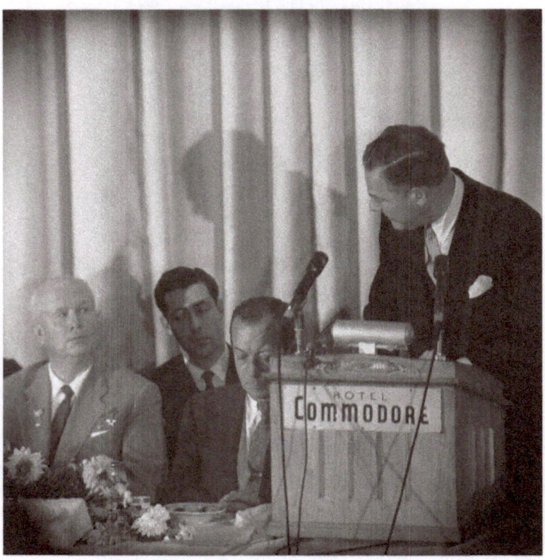

FIGURE 17.1 Nikita Khrushchev didn't need an interpreter for the expression of Henry Cabot Lodge, 18 September 1969

Khrushchev's son stressed the political importance of the visit as marking "the beginning of the mutual interaction of the two worlds" (Kordon, 2009). Four individuals are represented in the photograph: Henry Cabot Lodge, Nikita Khrushchev, Robert Wagner, and Khrushchev's interpreter, Viktor Sukhodrev (1932–2014). The photograph is accompanied by a caption that reads:

> Nikita Khrushchev didn't need an interpreter for the expression of Henry Cabot Lodge (right), US ambassador to the United Nations. Lodge took the rostrum in New York to chide the Russian premier for saying he saw no difference between old-time and modern US capitalism. They were at a luncheon with New York's mayor, Robert Wagner (next to Lodge). The other man is an interpreter.

Figure 17.2 portrays Edmund S. Glenn (1915–87) working as a booth interpreter. The photo is accompanied by a caption that reads:

> He turns Russian into English, but fast. Edmund S. Glenn, pictured wearing earphones of his trade, is the US State Department's chief interpreter. He has had the job of translating the speeches of Russia's Andrei Gromyko into English at the Japanese peace treaty conference. Those who know both languages say that Glenn's English translation is barely more than a word behind the flying Russian speeches of Gromyko. It is Glenn's voice which radio and television audiences across the nation have heard while Gromyko was speaking.

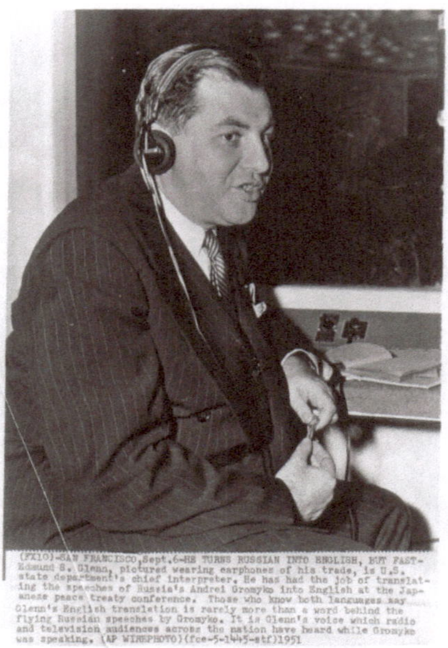

FIGURE 17.2 Edmund S. Glenn, 12 September 1951

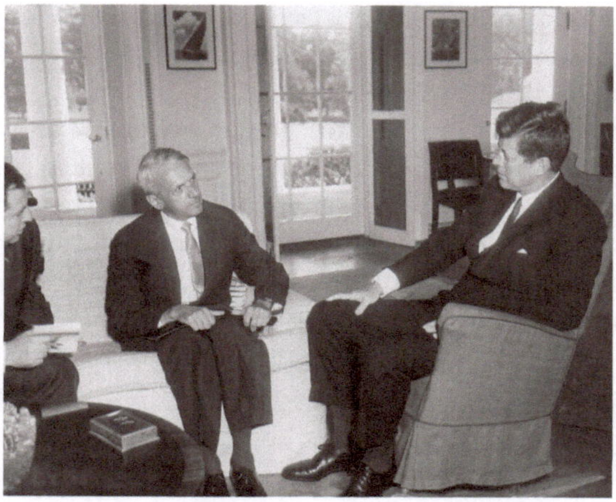

FIGURE 17.3 President John F. Kennedy meets with the Ambassador of the Polish People's Republic (Poland), Edward Drozniak, 19 July 1961

Finally, Figure 17.3 records a 1961 meeting between President Kennedy and the newly appointed Polish ambassador to the United States, Edward Drozniak (1902–66), with Glenn interpreting. Drozniak would serve as ambassador until his death in 1966, at the age of 64. Polish–American relations would be especially strained during this period due to the Cuban Missile Crisis, among other things. In 1963, Drozniak would refuse permission to American Catholic bishops to visit Poland to celebrate the one-thousandth anniversary of Christianity there. This photograph is accompanied by the caption: "President John F. Kennedy Meets with the Ambassador of the Polish People's Republic (Poland), Edward Drozniak (+ interpreter Edmund Glenn)".

Translating interpreters

Three post-positivist concepts will be central to my analysis of these photographs: Roman Jakobson's semiotic definition of translation, Heidegger's concept of enframing and Derrida's concept of the supplement.

Translation

Roman Jakobson opens his classic article of 1959 "On Linguistic Aspects of Translation" with a firm rebuttal of Bertrand Russell's positivist stance on the relationship of language to reality as apprehended through the senses. As Jakobson argues, we have never tasted ambrosia, but we have no problem understanding what is meant by the word. Jakobson expresses the view of linguistic relativists, for whom "the reality we know is itself constituted by language and is not, therefore, an unmediated presence that can be re-presented" (Thomas, 2000, p. 6). Jakobson goes on to expand the notion of translation, redefining it in semiotic terms as an exchange or substitution of signifiers that can take place within a single sign system (intralingual), across verbal sign systems (interlingual) or across different sign systems (intersemiotic) (Jakobson, 1959/2000, pp. 113–18).

In performing a verbal analysis of visual images, we are, to use Jakobson's terminology, offering an intersemiotic translation of an act of intersemiotic translation (the photographing of a verbal exchange) of an act of interlingual translation (the interpreted communicative event itself) – making this a translation of a translation of a translation. As such, the project of translating interpreters necessarily problematizes the positivist model of representation, which posits an anterior world that can be apprehended without translation/interpretation. Instead, it suggests there is no unmediated access to "the thing itself", that our "reality" is always already framed.

Enframing

In his post-positivist study of photography, McQuire invokes Heidegger's concept of enframing, as developed in the essay "The Question Concerning Technology". Insofar as technology establishes a certain relationship to the world, "it forms a mechanism of enframing, which, far from being a mere technique of performance, constitutes an essential part of the definition of the task to be performed" (McQuire, 1998, p. 4). Adapted to photography, the concept of enframing focuses attention on the ways in which photographs construct the reality they purport to reflect. Often ignored by the viewer, or rather perceived only at an unconscious level, enframing is perhaps the subtlest way in which the photographer shapes the viewer's perception *qua* interpretation of the scene that is framed.

We can also employ the concept of enframing in a more figurative sense by examining the ways that the captions accompanying the photographs, introduced not by the photographer but by a third party, "guide" the viewer's interpretation of the interpreted communicative event by suggesting which elements of the photograph are significant, or signifying, and which elements are not. Another figurative way to think of enframing is more specific to the context of interpretive communicative events: the role of the interpreter in "framing" the reception of the interpreted speech. Through a variety of linguistic and extra-linguistic means, from word choice to tone of voice and facial expressions, the interpreter frames the receiver's interpretation of the speaker's remarks.

Supplement

The Derridean supplement represents a post-positivist rereading of a concept defined by Rousseau as "an inessential extra added to something complete in itself". Derrida argues that one cannot add to something that is already complete in itself. For Derrida, the supplement becomes an addition that does not complete the thing but rather points to an originary lack, which calls forth endless supplementation.[3] The surplus of meaning in language, or polysemy, is a source of Derridean supplementarity, having the potential to unleash an endless chain of signification. Situations of conflict and stress have a tendency to activate these "supplemental" meanings in the speaker (as evidenced in the "Freudian slip"), as well as in the listener. Consider, for example, an incident that occurred during that same 1959 visit of the Soviet premier to the United States, when his remark "My vas pokhoronim" was rendered by his interpreter as "We will bury you." A reference to the final lines in Chapter 1 of Karl Marx's *Capital*, "What the bourgeoisie produces, above all else, is its own grave-diggers", the statement was interpreted by many Americans as a threat to use nuclear power against the United States. Strictly speaking, the interpreter's choice here cannot be described as a linguistic error. "To bury" is an acceptable translation of the Russian verb *pokhoronit'*, and can be found in any bilingual dictionary. It was the availability of other, supplemental meanings that caused the "mistranslation".

Another way to think of supplementarity is in relation to the captions that accompany photographs. Captions, one could argue, are necessarily supplemental, in the Derridean sense, insofar as they suggest a semantic lack in the photograph – this snapshot of "reality" must be supplemented by text. The converse, of course, is also true. The idea that news stories must be accompanied by visuals, in this case photographs, suggests that the story on its own is not self-sufficient.

Interpreting as enframing

I will discuss enframing first in the literal sense, as it relates to the photographs themselves, and then in a figurative sense, as it relates to the accompanying captions and the role of the interpreter. An important aspect of the framing of a photograph is the angle of the shot. For example, a dominant vector organizes the photograph in Figure 17.1, following the trajectory of Lodge's gaze down at the Soviet leader. In an attempt to solidify his credentials as an anti-communist, Lodge assumes an aggressive stance, turning from the audience to look directly at the Soviet premier. Moreover, Lodge, who is standing, appears to tower over the more diminutive Khrushchev seated nearby, an effect exaggerated by the angle of the photograph, and casts a menacing shadow over the interpreter. Similarly, in photograph 17.3 the interpreter is off to the side, while the two political figures look each other in the eye. In both photos 17.1 and 17.3, the "foreign" diplomat occupies the focal point. The viewer of the photograph would occupy a position similar to that of the US leaders – Lodge in Figure 17.1 and Kennedy in photograph 17.3. In the latter photograph, Drozniak is sandwiched between Glenn and Kennedy, and, like Khrushchev in photograph 17.1, is presented as an object of scrutiny, perhaps suspicion. This, one might argue, is a frame that undercuts the purported purpose of these meetings, which was to improve international understanding and cooperation.

We can also discuss framing in the literal sense, in terms of what or who is included within the frame – and, no less importantly, what or who is left out. Most obvious to a present-day observer, perhaps, is the fact that no women are present in the photographs, reinforcing the notion that politics, and by extension interpreting, was a man's business.[4] The idea of interpreting as a male domain is brought home in the subheading of White's article, which reads: "The disappearance of a universal diplomatic language of diplomacy makes *him* the indispensable *man* at today's conferences. The good ones are *towers of strength* in Babel" (1955, p. 12; italics added). Moreover, in all three photos the high-level interpreters are dressed in business suits – as opposed to military uniforms, which army interpreters wore during the Second World War – and so are barely distinguishable on a visual level from the politicians for whom they interpret, mystifying the question of the interpreter's political loyalty. The association of interpreting with modern technology – photograph 17.3 shows a composed Glenn alone in the interpreters' booth with a headset pressed against his ear – also suggests efficiency and objectivity, further underscored by the fact that Glenn is somewhat removed, literally and figuratively, from the interpreted event.[5] In photograph 17.3, Glenn holds a notepad, another professional tool of the interpreter.

The captions also frame these interpreted communicative events in ways that reveal a positivist bias. This is perhaps most evident in photograph 17.1, where the presentation of the event implies that the interpreter is occupying an ancillary role. For example, the opening sentence suggests that the interpreter's work is altogether unnecessary ("Khrushchev doesn't need an interpreter ... "), while the closing sentence mentions the interpreter but without providing his name – he is the only unnamed person in the photograph. The caption accompanying photograph 17.2 also implies a positivist notion of interpreting as the unproblematic replacement of the words of the speaker with those of the interpreter: Glenn's voice simply replaces that of Gromyko.

As mentioned above, the idea of framing can also be employed figuratively to describe the interpreter's work in shaping the reception of a politician's remarks. The Russian interpreter pictured in photograph 17.1 spoke very openly about this aspect of the interpreter's task in his memoir, *Iazyk moi – drug moi: Ot Khrushcheva do Gorbacheva* [*My Tongue Is My Friend: From Khrushchev to Gorbachev*] (Sukhodrev, 1999), and in many frank interviews he gave after the fall of the Soviet Union, in which the highly educated, cosmopolitan Sukhodrev described Khrushchev as "uneducated and uncultured" (Vandenko, 2013; my translation). The difference between Khrushchev and his interpreter in terms of education and culture, or class, is evident on a visual plane in their clothing (something captured especially well in the photographs of the time): the slim and stylish Sukhodrev is shown in well-tailored dark suits and sporting the latest haircut, in contrast to the Premier, who was short, slightly overweight and wore light-coloured, ill-fitting suits.[6] This difference in style and class was also evident in their speech. Khrushchev's speech was filled with folk sayings and colloquialisms, presenting major challenges for his interpreter (ibid.). Despite these differences, Khrushchev treated Sukhodrev with respect and kindness for, it seems, he understood that his interpreter made him look better to his foreign audience. As Sukhodrev puts it:

> For all of his denseness, Nikita Sergeevich understood how much he depended on the translator. He said one thing, but the foreign audience

heard, if you'll excuse me, something else. And so Khrushchev sized me up, felt me out. He gradually got used to me and accepted me.

(Ibid.; my translation)[7]

Interpreting as supplement

Supplementarity haunts photograph 17.1 in the shape of two people who are "outside" the immediate interpretive exchange of Lodge addressing Khrushchev. I am referring here first to the New York City mayor, Robert Wagner, the host of the event, who is caught "in between" Khrushchev and Lodge. Wagner looks down at the table, his hands folded in front of him, while Lodge confronts the Soviet premier. Wagner was a progressive Democrat who was then serving the second of three terms in office, while Lodge was a conservative Republican who one year later would be his party's nominee for president and would lose a close election to John F. Kennedy. The presence of Wagner in the photo complicates the idea that this is a simple bilateral exchange between Lodge and Khrushchev by suggesting a second audience for Lodge's remarks: the US electorate. Lodge was no doubt taking advantage of a well-publicized opportunity to present himself to the American people as a defiant Cold Warrior. The caption, however, makes no mention of the political affiliation of Lodge and Wagner, encouraging viewers to read the event in Cold War terms as Americans vs Russians.

The presence of the second supplemental figure in the photograph – the interpreter – also underscores the mediated nature of the exchange, although the caption writer plays this down by refusing to name the interpreter, suggesting that he is incidental to the proceedings. By withholding the name of the interpreter and the party affiliation of the mayor, the caption portrays these two men as Rousseauian supplements – that is, innocent additions to a thing complete in itself – in the exchange between Lodge and Khrushchev. This is also suggested in the caption to photograph 17.3, in which the presence of the interpreter is designated in parentheses with a plus sign and his name. The threat of the supplement is also defused in the positioning of the interpreter at the back (photograph 17.1) and off to the side (photograph 17.3), and in the headline of White's article, which mimics a mathematical formula: "Interpreter: Linguist plus Diplomat" (1955, p. 12).

Although the captions encourage us to read the interpreter as a Rousseauian supplement, the interpreters themselves, interviewed by White, reveal their activity to be something much closer to a Derridean supplement. The same interpreter who claimed that the interpreter's task was to be "absolutely faithful" to the speaker goes on to comment: "You keep up with him, but you also try to make things clear where they are muddy. Sometimes you interpret exactly what he says and then *add* a few words to explain what he means" (p. 12; emphasis added). But, if White's interpreter giveth, he also taketh away, admitting to removing "unnecessary" verbiage: "You do an editing job as you go along" (p. 12). These "additions" and "excisions" expose the originary lack in the source utterance, the fact that meaning does not exist *a priori* but is constructed in the interpretation, so that even the most "exact" linguistic rendering may require the interpreter to add things and to take things away. In other words, the speaker's "original" remarks must – as all language must – be subjected to interpretation, in the broader sense of the word, that is, they must be assigned meaning by the interpreter before they can be interpreted, in the narrow sense. These additions and excisions, noted so nonchalantly by White's interpreter, complicate the notion of semantic equivalence by highlighting interpreting as interpretation rather than as a simple matching of words.

Conclusion

If these photographs of Cold War interpreted events suggest, at first glance, the celebration of mimesis and the triumph of positivism, a close reading suggests that they presage the unravelling of those very concepts. If the positivist model of communication constructed the interpreter as a kind of linguistic machine, the pressures of the Cold War era exposed the need for the interpreter to assume a role with far greater agency – to become, as White suggests, a diplomat as well as a linguist. This seemingly innocent "addition" to the interpreter's skill set, however, theoretically undermines the positivist model of representation. Absolute fidelity to the "original" utterance is now tempered with a concern over how the message is received, which is another kind of fidelity altogether. This necessarily complicates the notion of interpreters as neutral conduits, presenting them instead as co-constructors of meaning, framers of the event. Soon after these photos were published, attempts to accommodate these two aspects of the interpreter's role would come to preoccupy scholars working in the emerging field of translation studies, forcing a rethinking of interpreter ethics that is still ongoing today.

Notes

1 Glenn went on to publish more works on translation and cross-cultural communication, most notably *Man and Mankind: Conflict and Communication between Cultures* (1981), co-authored with Christine G. Glenn. After retiring from the Language Services Division of the US Department of State, Glenn became a professor of international communications at the University of Delaware.

2 The Cold War is often defined as the period 1947–91.

3 The concept of the supplement and supplementarity were developed in Derrida's seminal work *On Grammatology*, where he describes the supplement as both "accretion and substation" (1977, p. 144).

4 Although a woman, Nora Lejins, was in fact the head of language services in the US State Department at that time (see Baigorri-Jalón in this volume); the number of women among the ranks of high-echelon interpreters was very small.

5 Glenn is also shown fidgeting with keys, probably to the booth, which is no doubt kept locked to protect the expensive new technology.

6 It is perhaps no surprise that Sukhodrev was completely enamoured of President Kennedy, whom he met at the Vienna talks in 1961. Sukhodrev appeared to identify more closely with the US president than with the Soviet premier. As he put it: "Of course, John Kennedy made an indelible impression on me. If you want to explain what the word

'charisma' means, it was Kennedy! … Kennedy literally exuded magnetism and charm" (Sharalapova and Poliakova, 2013).

7 This is a situation very different from the one in China, where Mao was deeply suspicious of his highly educated interpreters and repeatedly purged them (Baigorri-Jalón and Fernández-Sánchez, 2010, p. 9).

References

Baigorri-Jalón, J. and Fernández-Sánchez, M. M., 2010. *Understanding High-level Interpreting in the Cold War: Preliminary Notes*. Paris: Presses de la Sorbonne Nouvelle.

Derrida, J., 1977. *On Grammatology*. Translated by G. Ch. Spivak. Baltimore, MD: The Johns Hopkins University Press.

Glenn, E. S., 1955. *Language and Patterns of Thought*. Washington, DC: School of Foreign Service, Georgetown University.

Glenn, E. S. and Glenn, C. G., 1981. *Man and Mankind: Conflict and Communication between Cultures*. Norwood, NJ: Ablex.

Jakobson, R., 1959/2000. "On linguistic aspects of translation." In: L. Venuti, ed., 2000. *The Translation Studies Reader*. London: Routledge. pp. 113–18.

Kordon, K., 2009. "Krushchev comes to America: the advent of mutual understanding." *Voces Novae. Chapman University Historical Review*, 1(1). http://journals.chapman.edu/ojs/index.php/VocesNovae/article/view/19/65 [accessed 7 April 2014].

McQuire, S., 1998. *Visions of Modernity: Representation, Memory, Time and Space in the Age of the Camera*. London: Sage.

Raley, R., 2003. "Machine translation and global English." *Yale Journal of Criticism*, 16(2), pp. 291–313.

Sharalapova, E. and Poliakova, A., 2013. "Legendy perevodcheskogo fronta: Viktor Sukhodrev" ["Legends from the translation front: Viktor Sukhodrev"]. [Interview with Viktor Sukhodrev]. In: *Legendy perevodcheskogo fronta* [*Legends from the Translation Front*]. www.okperevod.ru/pg-3.htm [accessed 7 April 2014].

Sukhodrev, V., 1999. *Iazyk moi – drug moi: ot Khrushcheva do Gorbacheva* [*My tongue – my friend: From Khrushchev to Gorbachev*]. Moscow: ACT.

Thomas, J., 2000. "Introduction." In: J. Thomas, ed. *Reading Images*. London: Palgrave. pp. 1–10.

Vandenko, V., 2013. "Ne vse zoloto chto molchit" ["Not everything that's silent is golden"]. [Interview with Viktor Sukhodrev]. In: *Liudi. Peoples.ru*. www.peoples.ru/state/citizen/suhodrev/ [accessed 7 April 2014].

White, P. T., 1955. "The interpreter: linguist plus diplomat." *New York Times Magazine*, 6 Nov. CV(35,715), pp. 12, 30, 32–33.

INDEX